AF780234

Applications Enabled by FPGA-Based Technology

Applications Enabled by FPGA-Based Technology

Editors

Andres Upegui
Andrea Guerrieri
Laurent Gantel

MDPI

Basel • Beijing • Wuhan • Barcelona • Belgrade • Novi Sad • Cluj • Manchester

Editors
Andres Upegui
University of Applied Sciences of Western Switzerland
Geneva, Switzerland

Andrea Guerrieri
University of Applied Sciences of Western Switzerland
Sion, Switzerland

Laurent Gantel
University of Applied Sciences of Western Switzerland
Geneva, Switzerland

Editorial Office
MDPI
St. Alban-Anlage 66
4052 Basel, Switzerland

This is a reprint of articles from the Special Issue published online in the open access journal *Electronics* (ISSN 2079-9292) (available at: https://www.mdpi.com/journal/electronics/special_issues/FPGA_electronics).

For citation purposes, cite each article independently as indicated on the article page online and as indicated below:

Lastname, A.A.; Lastname, B.B. Article Title. *Journal Name* **Year**, *Volume Number*, Page Range.

ISBN 978-3-0365-8784-4 (Hbk)
ISBN 978-3-0365-8785-1 (PDF)
doi.org/10.3390/books978-3-0365-8785-1

Cover image courtesy of Andres Upegui, Andrea Guerrieri and Laurent Gantel

© 2023 by the authors. Articles in this book are Open Access and distributed under the Creative Commons Attribution (CC BY) license. The book as a whole is distributed by MDPI under the terms and conditions of the Creative Commons Attribution-NonCommercial-NoDerivs (CC BY-NC-ND) license.

Contents

About the Editors vii

Sunil Puranik, Mahesh Barve, Swapnil Rodi and Rajendra Patrikar
FPGA-Based High-Throughput Key-Value Store Using Hashing and B-Tree for Securities Trading System
Reprinted from: *Electronics* **2023**, *12*, 183, doi:10.3390/electronics12010183 . . . . **1**

Sunil Puranik, Mahesh Barve, Swapnil Rodi and Rajendra Patrikar
Acceleration of Trading System Back End with FPGAs Using High-Level Synthesis Flow
Reprinted from: *Electronics* **2023**, *12*, 520, doi:10.3390/electronics12030520 . . . . **19**

Grzegorz Jabłoński, Piotr Amrozik and Krzysztof Hałagan
A Model of Thermally Activated Molecular Transport: Implementation in a Massive FPGA Cluster
Reprinted from: *Electronics* **2023**, *12*, 1198, doi:10.3390/electronics12051198 . . . . **35**

Hyun-Sik Choi
Electromyogram (EMG) Signal Classification Based on Light-Weight Neural Network with FPGAs for Wearable Application
Reprinted from: *Electronics* **2023**, *12*, 1398, doi:10.3390/electronics12061398 . . . . **47**

Fanpeng Kong, Manuel Cegarra Polo and Andrew Lambert
FPGA Implementation of Shack–Hartmann Wavefront Sensing Using Stream-Based Center of Gravity Method for Centroid Estimation
Reprinted from: *Electronics* **2023**, *12*, 1714, doi:10.3390/electronics12071714 . . . . **61**

Ning Mao, Haigang Yang and Zhihong Huang
An Instruction-Driven Batch-Based High-Performance Resource-Efficient LSTM Accelerator on FPGA
Reprinted from: *Electronics* **2023**, *12*, 1731, doi:10.3390/electronics12071731 . . . . **81**

Lukáš Kohútka and Ján Mach
A New FPGA-Based Task Scheduler for Real-Time Systems
Reprinted from: *Electronics* **2023**, *12*, 1870, doi:10.3390/electronics12081870 . . . . **97**

Jonathan Cureño-Osornio, Israel Zamudio-Ramirez, Luis Morales-Velazquez, Arturo Yosimar Jaen-Cuellar, Roque Alfredo Osornio-Rios and Jose Alfonso Antonino-Daviu
FPGA-Flux Proprietary System for Online Detection of Outer Race Faults in Bearings
Reprinted from: *Electronics* **2023**, *12*, 1924, doi:10.3390/electronics12081924 . . . . **115**

Ali Ebrahim
Finding the Top-K Heavy Hitters in Data Streams: A Reconfigurable Accelerator Based on an FPGA-Optimized Algorithm
Reprinted from: *Electronics* **2023**, *12*, 2376, doi:10.3390/electronics12112376 . . . . **133**

Adrien Bourennane, Camel Tanougast, Camille Diou and Jean Gorse
Accurate Multi-Channel QCM Sensor Measurement Enabled by FPGA-Based Embedded System Using GPS
Reprinted from: *Electronics* **2023**, *12*, 2666, doi:10.3390/electronics12122666 . . . . **155**

Le Yu, Baojin Guo, Tian Zhi and Lida Bai
Improving Seed-Based FPGA Packing with Indirect Connection for Realization of Neural Networks
Reprinted from: *Electronics* **2023**, *12*, 2691, doi:10.3390/electronics12122691 . . . . **169**

Andrea Guerrieri, Andres Upegui and Laurent Gantel
Applications Enabled by FPGA-Based Technology
Reprinted from: *Electronics* **2023**, *12*, 3302, doi:10.3390/electronics12153302 . . . . **183**

About the Editors

Andres Upegui

Andres Upegui (Associate Professor) graduated in Electronic Engineering from the Universidad Pontificia Bolivariana, Medellín, Colombia in 2000. In the same year he joined the group of Micrelectronics in the same university. From 2001 to 2002 he followed the Graduate School on Computer Science at the EPFL, and then he joined the Logic Systems Laboratory (LSL) as PhD student. In 2006, he received the title of PhD. from the EPFL for his thesis entitled "Dynamically reconfigurable bio-inspired hardware". Afterwards he worked as senior researcher and lecturer at the HEIG-VD, Yverdon and Hepia, Geneva. Since 2011 he is Professor in embedded systems at Hepia, Geneva, Switzerland, he is co-head to the CoRES group (Communicating Reconfigurable Embedded Systems) and founding member of the AI Swiss Center for SMEs. His research interest include computer architectures, reconfigurable computing, self-adapting hardware systems, and embedded machine learning.. . . .

Andrea Guerrieri

Andrea Guerrieri is Associate Professor and Researcher at the University of Applied Science Western Switzerland (HES-SO) and École Polytechnique Fédérale de Lausanne (EPFL). He graduated in electronic engineering from the Politecnico di Torino, Turin, Italy, in 2015. In 2017, he joined the Processor Architecture Laboratory, EPFL, Switzerland, and from 2020 he is also affiliated with the HES-SO. His research interest includes electronic design automation, reconfigurable computing, and security. Recent projects involve high-level synthesis, reconfigurable SoCs exploiting dynamic partial reconfiguration of FPGAs for future space missions, exoplanet observation, and post-quantum cryptography. He is a co-developer of the dynamically scheduled high-level synthesis compiler Dynamatic, and co-author of a book on system-on-chip design with Arm. He is a Senior Member of IEEE, reviewer, and artifact evaluator for scientific conferences and journals such as ISCAS, FCCM, and FPGA. He also serves as a technical program committee member for DAC and ICCD, and Swissuniversities for the Open Science Programme. Prof. Guerrieri is also the recipient of the Best Paper Award at FPGA'20, held in Seaside, California, Best Paper Nominations at FCCM'22, held in New York, USA, FPL'22 held in Belfast, UK.

Laurent Gantel

Laurent Gantel (Ph.D) received a Research Master's degree in Autonomous Systems Engineering at the National School of Electronics and its Applications (ENSEA - France), in 2009, and a Ph.D degree in from the CY Cergy Paris University (France) in 2014. In 2013, he joined the University of Applied Sciences and Arts of Western Switzerland (HES-SO), in Geneva, and actively participated on various research projects related to FPGA-based heterogeneous platforms, high-speed communication, and cyber-physical systems. Between 2016 and 2018, he worked as a design and development engineer at the Annecy le Vieux Particle Physics Laboratory (LAPP - France), and collaborated on the ATLAS experiment conducted at CERN, in Geneva. He is currently a Scientific Assistant at the HES-SO and his research interests mainly concern FPGA acceleration, dynamically reconfigurable System-on-Chip, signal processing, smart cities, and machine learning inference for FPGAs and low-power embedded devices.

electronics MDPI

Article

FPGA-Based High-Throughput Key-Value Store Using Hashing and B-Tree for Securities Trading System

Sunil Puranik [1,*], Mahesh Barve [1], Swapnil Rodi [1] and Rajendra Patrikar [2]

[1] Tata Consultancy Services, Pune 411057, India
[2] Department of Electronics, Visvesvaraya National Institute of Technology, Nagpur 440012, India
* Correspondence: sunilsavitap@students.vnit.ac.in

Abstract: Field-Programmable Array (FPGA) technology is extensively used in Finance. This paper describes a high-throughput key-value store (KVS) for securities trading system applications using an FPGA. The design uses a combination of hashing and B-Tree techniques and supports a large number of keys (40 million) as required by the Trading System. We have used a novel technique of using buckets of different capacities to reduce the amount of Block-RAM (BRAM) and perform a high-speed lookup. The design uses high-bandwidth-memory (HBM), an On-chip memory available in Virtex Ultrascale+ FPGAs to support a large number of keys. Another feature of this design is the replication of the database and lookup logic to increase the overall throughput. By implementing multiple lookup engines in parallel and replicating the database, we could achieve high throughput (up to 6.32 million search operations/second) as specified by our client, which is a major stock exchange. The design has been implemented with a combination of Verilog and high-level-synthesis (HLS) flow to reduce the implementation time.

Keywords: key-value store; Field-Programmable Gate Arrays (FPGA); high-level synthesis (HLS); system performance; B-Tree; hashing; CAM-content addressable memory; high-bandwidth memory (HBM); Block-RAM (BRAM)

Citation: Puranik, S.; Barve, M.; Rodi, S.; Patrikar, R. FPGA-Based High-Throughput Key-Value Store Using Hashing and B-Tree for Securities Trading System. *Electronics* **2023**, *12*, 183. https://doi.org/10.3390/electronics12010183

Academic Editor: Akash Kumar

Received: 9 December 2022
Revised: 22 December 2022
Accepted: 23 December 2022
Published: 30 December 2022

Copyright: © 2022 by the authors. Licensee MDPI, Basel, Switzerland. This article is an open access article distributed under the terms and conditions of the Creative Commons Attribution (CC BY) license (https://creativecommons.org/licenses/by/4.0/).

1. Introduction

Key-value search is one of the most basic operations of data processing. Key-value search has a wide range of applications such as forwarding table lookup in routers, directory search, data deduplication in storage, stock trading, and graph search. Most of these operations require line rate processing, so low latency is an important requirement for the key-value search.

Field-Programmable Gate Arrays (FPGAs) [1] are programmable digital integrated circuits that allow a hardware designer to program the customized digital logic as per the requirement. FPGAs are widely used in the field of financial processing. A well-known example of the application of FPGA in financial systems is the deployment of FPGAs in the London Stock Exchange for securities trading [2]. FPGAs support dynamic reconfiguration in contrast to ASICS, which is a very important feature in the design of trading systems as frequent changes in the algorithms are required. Furthermore, the number of stock exchanges is not large enough to justify the use of ASIC technology, which will be costly when the volumes are not very high.

In this paper, we describe the design of a key-value store (KVS) for a securities trading system that needs to support a very large number of keys (up to 40 million). Since the KVS is used for a securities-trading application, its functionality and specifications are different from a conventional KVS. KVS systems typically support search, insert, and delete operations. In a search operation, a key is input into KVS, and data corresponding to the key is retrieved. In an insert operation, a new key along with the data is written into the database, while in a delete operation, the key and the corresponding data are removed from the database. We list the specifications of our KVS and the differences between the functionality of our KVS and the conventional KVS. These specifications have been received from a major stock exchange, which is our client:

1. KVS has to support 40 million keys. This is based on the number of clients the stock exchange has to handle. There is one key for each client, and the number of clients is 20 million and is expected to grow to 40 million in the future.
2. The key length is 128 bits.
3. Key length is fixed, and KVS does not support variable length keys.
4. KVS supports search, insert, and delete key operations.
5. The throughput of the KVS should be 4 million search operations/s (the order processing rate of the stock exchange is 2 million orders/s, which is expected to increase to 4 million orders/sec in the future. One order generates one search request).
6. Data associated with the key are 128 bits.
7. Insert and delete operations are performed in bulk, and unless insert and delete operations are completed, the corresponding key is not searched for.
8. Insert and delete operations performed in bulk should not affect search operations. It should be possible to perform search operations in parallel with bulk insert and delete operations. Search operations need to be carried out at the line rate.

It can be seen from above that the features of our KVS are rather different from those of a conventional KVS [3]. In a conventional KVS, insert and delete operations are dynamic and typically not performed in bulk. While insert and delete operations are being performed, search operations cannot be performed in parallel. Furthermore, in a conventional KVS, if insert and search operations are pipelined, there is a possibility of a read-after-write (RAW) hazard if the insert operation is not completed (committed to the memory) before the search operation. This situation does not arise in our KVS since the search operation for the newly inserted key will not be performed until the insert is completed, i.e., insert data are committed to memory. Newly inserted keys correspond to the new clients added to the stock exchange, and these clients will not start sending the trade requests (which will generate search operations) until the database is completely written to the memory. New clients and their data are, however, added in bulk as stated above in the specifications.

This paper is organized as follows—in Section 2, we describe the related work. In Section 3, we describe the operation of the trading system and the role of KVS in a trading system. Section 4 discusses the architecture of KVS in detail and Section 5 discusses simulation results. We conclude the paper in Section 6 regarding future work.

2. Related Work

Key-value stores have been traditionally implemented with x86 servers [4,5], but they are not optimized for this kind of workload. The processor's last-level data cache is not effectively used due to the random-access nature and memory size of the application. Furthermore, the throughput and latency are affected by the high latency of the TCP/IP stack implemented with the software. Therefore, these key-value stores are not suitable for financial and trading systems where line rate processing is required and latencies have to be low. With the FPGA technology, we can implement data flow architectures, and instruction- and task-level parallelism in FPGAs can be used to significantly increase the throughput and reduce the latency—an important requirement for line rate processing.

Key-value (KVS) stores can be implemented in FPGA logic using techniques such as hash tables, CAMs, and B-Trees [6–9]. The main challenge with the hashing technique is handling the case when multiple keys map to the same hash index (which is called a *hash collision*) while maintaining consistent throughput levels. One well-known approach to avoid collisions completely is to use *perfect hashing* [10]. If all the input keys are previously known, collisions can be avoided using a customized hash function. The key benefit of this approach is that all search operations can be performed in O (1) time. This approach is not, however, suitable for trading systems that use cases as the input keys are dynamic and previously unknown. The keys continue being added in bulk as new customers are registered in a trading system. Another technique to handle the collisions is to store the colliding keys in a dynamically linked list often referred to as *a bucket*, which is pointed to by a hash index. However, keys need to be accessed sequentially and compared with the input key, which increases the time for the search operation. We can allocate fixed space to store the colliding keys contiguously in a bucket so that the keys can be read in a burst from off-chip memory. The keys read from memory can be compared in parallel in FPGA logic by using multiple comparators operating in parallel. This approach is used in [11]. However, allocating a fixed storage space results in non-optimal memory utilization if a small number of keys map to the particular hash index. This is especially true if we are using an on-chip BRAM to store the colliding keys. To overcome this problem and conserve BRAM (which is a scarce resource in the FPGA), our design uses a novel technique of hashing with buckets of different capacities (by different we mean buckets of pre-determined sizes of 1, 4, 8, and 16) as explained later.

In a trading systems application, 90% of the search requests are generated by 5% of users referred to as *active users*. We keep the keys and the corresponding data values for up to 1 million active users in BRAM/UltraRAM. The actual number of keys stored in BRAM/UltraRAM will depend on the collision pattern. This is to increase the overall throughput of KVS. The data of approximately 39 million users is kept in on-chip high-bandwidth memory (HBM). We have a very large amount of HBM—16 GBytes available in the FPGA, which we are using for implementing the KVS. While we are using a hash-based lookup engine for keys stored in BRAM, we are using a B-Tree-based lookup engine for keys stored in the HBM. An alternative to B-Tree is to use the cuckoo hashing technique [12–14] in which items are stored in one of the two possible locations. With cuckoo hashing, search operations take a constant time; however, insert operations pay the cost of collisions. Inserting a new key can result in 'cycles', and in the case of a cycle, new hash functions are chosen and the whole data structure needs to be 'rehashed'. Multiple rehashes might be necessary before Cuckoo succeeds. This results in very high response times for insert operations. Consequently, this approach becomes unsuitable for our application, considering the large number of insert operations performed in bulk. To simplify insert operations, we use the B-Tree approach in which insert and delete operations will be handled in software while the search operations will be executed in hardware. Because of the above-mentioned points, we use hashing for implementing lookups in BRAM, which can store a relatively small number of keys, and B-Tree for implementing lookups in HBM, which can handle a very large number of keys.

3. Operation of a Trading System

The architecture of a Trading System is depicted in Figure 1. It consists of many users/traders connected to the trading system using an IP-based network. These traders are outside the premise of the trading system.

The Trading System itself consists of front-end, order-matching, and post-trade components. The front-end component receives trade requests from the users on the Ethernet network and performs validations on the ranges of various parameters in a trade request and the functional validations. The order-matching component matches buy orders against sale orders and vice versa. It maintains the database of orders received from traders and executes commands to perform order matching. The post-trade block performs data logging functions after a trade happens and also performs the functions of record keeping, journaling, and sending a response back to the users. Since these functions are implemented in software at present in most trading systems, these systems give a high response time and low throughput for trade requests.

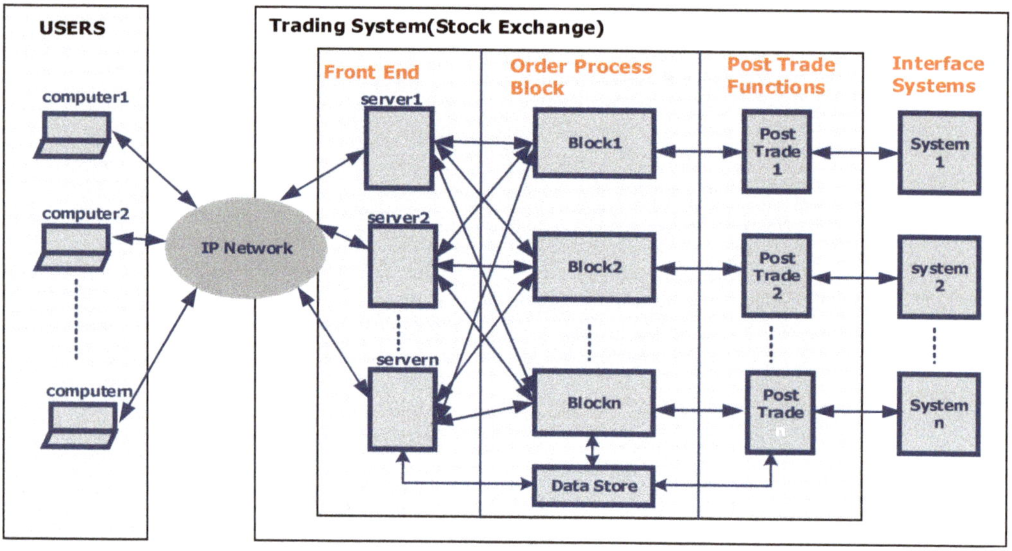

Figure 1. Architecture of a Trading System.

FPGAs are increasingly being used to perform these functions in hardware to reduce latency and increase throughput [15,16]. Front-end and order-matching functions are implemented with PCIe add-on FPGA boards. As mentioned above, the front-end functions involve validations of the numeric ranges of different fields present in the trade requests and also the functional validations. KVS is used mainly for functional validations. An example of functional validation is using a nine-digit social security number to retrieve the identification number (ID) of the user and compare it with the user ID in the database to check if a particular user is a valid user. Since there are approximately 20 million users in our client stock exchange at present and this number is expected to grow to 40 million in the future, our KVS design needs to handle 40 million keys.

Moreover, at present, the rate at which orders are received is 1 million orders/s and this is expected to grow to 4 million orders/s in the future. Therefore, the KVS needs to support up to 4 million search operations/s.

4. Architecture of KVS_Top

KVS_top is our top-level design, which instantiates both the hash-based lookup engine and B-Tree-based lookup engines. Since our KVS_Top design uses both the BRAM and HBM as stated earlier, the design is partitioned into two main sub-blocks, KVS_bram, which stores the keys in BRAM/UltraRAM, and KVS_hbm, which stores keys in on-chip HBM. KVS_bram executes insert, delete, and search commands in hardware. KVS_hbm executes search commands in hardware while insert and delete commands are issued to KVS_hbm through software by the host. KVS_bram uses hash tables while KVS_hbm uses four lookup engines working in parallel on four B-Tree databases. Four B-Tree databases contain the same data, as explained later.

The top-level block diagram of KVS_Top showing all the interfaces is shown in Figure 2. KVS_Top consists of KVS_hbm and KVS_bram as stated earlier. Search commands are issued to KVS_Top from users submitting trade requests on the Search_Commands_Interface. A search command consists of a Command_code (2-bit) and an input key (128-bit).

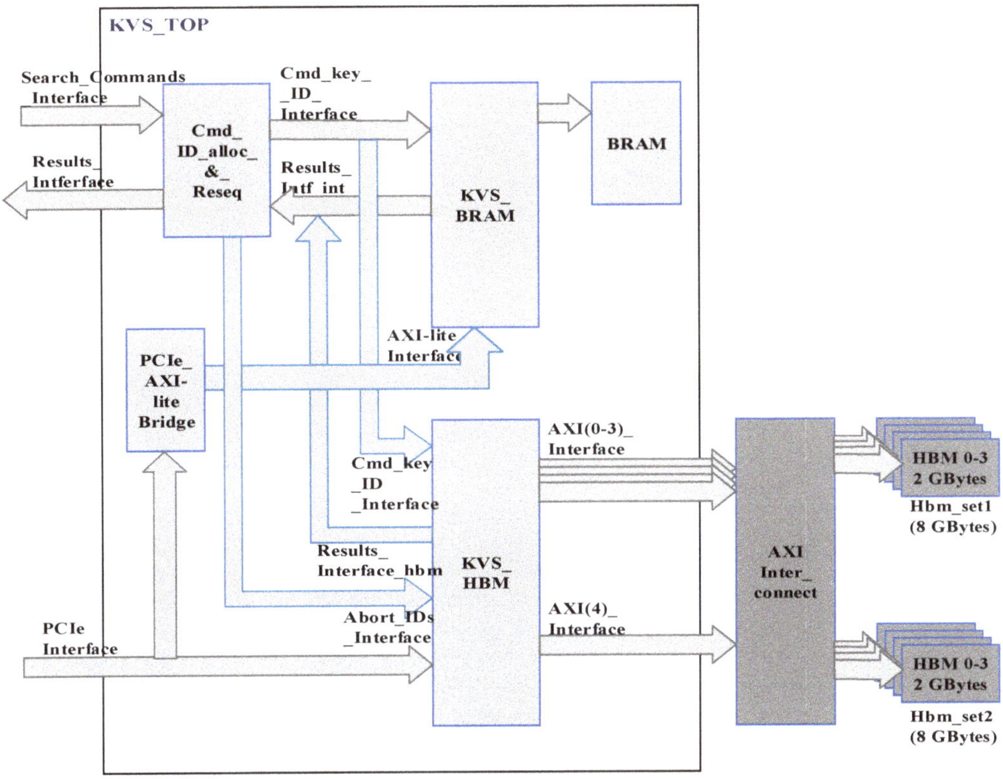

Figure 2. Block Diagram of KVS_Top showing all Interfaces.

Insert and delete commands are issued to KVS_bram and KVS_hbm through the software on the PCIe interface shown in Figure 2. The PCIe_Axi-lite bridge converts the PCIe protocol to AXI-lite for interfacing with KVS_bram. KVS_bram accepts the insert and delete commands on the AXI-lite interface. The Cmd_ID_alloc_&_reseq block allocates a 2-bit Command_ID to each search command. Search commands are supplied to both KVS_bram and KVS_hbm in parallel since it is not known in advance whether a key resides in BRAM or HBM. The search command execution time of KVS_bram is much shorter than that of KVS_hbm. Therefore, if KVS_bram completes the search command successfully as indicated on the internal results interface—Results_intf_int—the Cmd_ID_alloc_&_reseq block commands the KVS_hbm to abort the search command. The abort command is sent to KVS_hbm on the Abort_IDs_Interface by issuing Abort_ID, which is equal to the Command_ID of the command to be aborted, as shown in Figure 2. If the search command is completed with an error status from KVS_bram on the internal results interface, Results_Intf_int, KVS_hbm is allowed to complete the command. The Command_IDs are necessary since the search commands given to KVS_hbm may not be completed in the same sequence in which they are issued. For example, if two search commands, cmd1 and cmd2, are submitted to KVS_hbm in the pipelined manner and the key corresponding to cmd1 is found in level 3 of B-Tree and the key corresponding to cmd2 is found in level 1 of B-Tree, then cmd2 will be completed before cmd1. The Cmd_ID_alloc_&_reseq block re-sequences the command responses so that the results of search command execution are received on the external Results_Interface in the same sequence in which search commands are issued to KVS_Top. As stated earlier, KVS_bram stores the keys in on-chip BRAM, and KVS_hbm stores the keys in BRAM and on-chip HBM. There is 16 GB of on-chip HBM available in the FPGA, which we use for KVS. Sixteen gigabytes of HBM are divided into two sets (hbm_set1 and hbm_set2) of four banks with each bank consisting of 2 GBytes of memory (Refer to Figure 2). Four banks of HBM in one set are accessed by four lookup engines present in KVS_hbm, as explained later. The key database is replicated in all four banks in one set while inserting the keys using the software. Therefore, all four lookup engines work in parallel, giving very high throughput. Since the two sets of banks are connected using an on-chip AXI-Interconnect, each set can be accessed by four lookup engines. When hbm_set1consisting of four banks of 2Gbytes each is accessed by four lookup engines for carrying out search commands, the software writes the insert and delete commands in bulk to hbm_set2 and prepares an alternate database. This is performed in parallel while search operations are carried out by four lookup engines on hbm_set1. When an alternate database is ready, the software sets a flag called bank_flag in a register inside KVS_hbmm and four lookup engines then switch to hbm_set2 to carry out search operations and hbm_set1 will be used to carry out the new insert and delete operations in the software. KVS_hbm accesses two sets of HBM using AXI (0–3) interfaces for search commands and the AXI (4) Interface shown for insert and delete commands issued through software (Refer to Figure 2). It can access both sets of HBM simultaneously due to the presence of AXI-Interconnect. The architecture of KVS_bram and KVS_hbm is explained in detail below.

4.1. KVS_Bram

KVS_bram completely reuses the architecture of the lookup engine designed in our earlier work [17] and uses a hash table and buckets of different capacities as mentioned earlier. In a traditional KVS, which uses hashing, colliding keys are stored in a linked list (called a bucket) pointed to by a hash index. The keys in a linked list need to be sequentially read and compared with the incoming key. Though the usage of a linked list conserves BRAM, it increases search latency and reduces the throughput if the number of colliding keys is very large. To overcome this, our design uses buckets of variable capacity to conserve the BRAM and stores the keys contiguously in a bucket. These keys are read into a pipelined comparator and compared in parallel with an incoming search key to reduce the search latency.

In our design, the 128-bit incoming key is hashed using the checksum algorithm and the resulting hash value is used as an index into the hash table (implemented in BRAM), which is used to store bucket descriptors (Bucket_dsc0—Bucket_dscn-1) as shown in Figure 3. Bucket descriptors in turn contain pointers to the buckets of different capacities containing a differing number of keys. The bucket descriptor structure contains the key count (Key_cnt), pointer valid bit (Ptr_valid), pointer type (Ptr_type), offset of the bucket group (Offset), and bucket number (Bucket_no). The description of these fields is given below:

1. Ptr_valid (1 bit)—Indicates contents of bucket descriptor are valid and bucket descriptor holds a valid pointer to a hash bucket containing one or more keys.
2. Ptr_type (2 bits)—Indicates the type of hash bucket or capacity of the hash bucket in Block RAM as explained below. Bit 00, 01, 10, and 11 indicate 1-key, 4-key, 8-key, and 16-key buckets, respectively.
3. Bucket_ptr (16 bits+ 4 bits)—20-bit pointer to hash bucket pointed to by this descriptor as shown in Figure 3. Bucket_ptr consists of a 16-bit Offset to Bucket Group and 4-bit Bucket_no.
4. Key_cnt (5 bits)—Indicates the count of colliding keys in the hash bucket pointed to by the pointer in this descriptor.

As mentioned above, our KVS design uses buckets of four different sizes to conserve the BRAM, instead of using buckets of fixed lengths. Using buckets of fixed lengths results in wastage of BRAM if the number of colliding keys in the bucket is small. If initially there is only one key in the hash bucket, a bucket size of 1 is used. If one more key hashes into the same bucket during the insert operation, a bucket with a size of 4 is obtained from the free pool of buckets, and the original 1-key bucket is returned to the free pool of buckets. If a 4-key bucket is full and one more key hashes into it, the 4-key bucket is returned to the free pool and an 8-key bucket is obtained from the free pool, and all keys are transferred from the 4-key bucket to the 8-key bucket. Similar logic holds for the 8-key and 16-key buckets. The free pool of buckets is a First-In-First-Out (FIFO), which is initialized at the reset with the pointers to the free buckets. Since buckets are stored in BRAM, which is limited in FPGA, this scheme optimizes the use of BRAM.

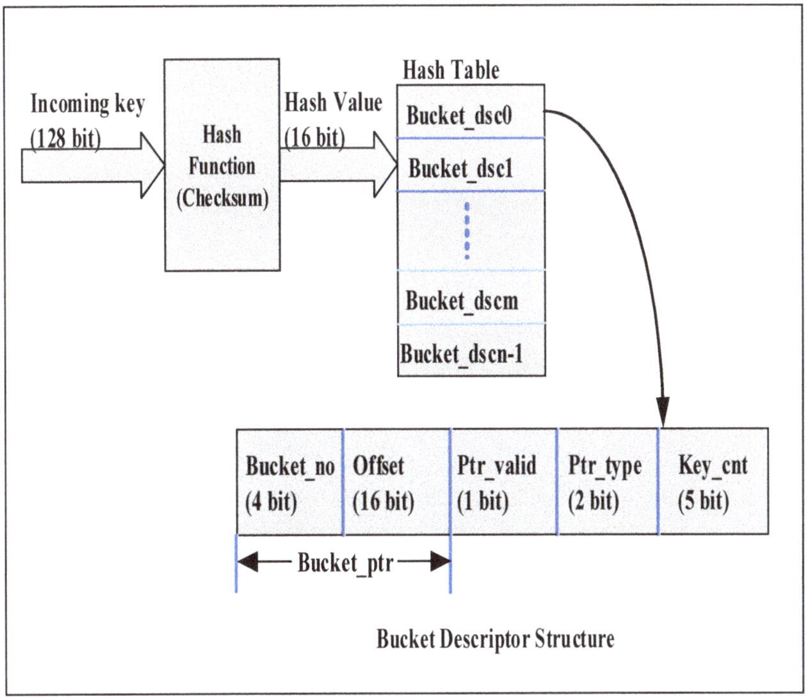

Figure 3. Hash Table and Bucket Descriptor Data Structure.

Keys are stored contiguously in a hash bucket, and using a pointer in the bucket descriptor, all keys in the hash bucket are read into a parallel comparator in FPGA. The comparator compares all the keys in the hash bucket in parallel with the incoming key and returns the index of the matching key, which is used to access the data corresponding to that key. Parallel comparison gives very low latency, and the pipelined operations of different blocks result in very high throughput.

Hash buckets are stored in bucket memory in BRAM, which consists of an array of bucket groups. As shown in Figure 4, each bucket group can consist of one 16-key bucket, two 8-key buckets, four 4-key buckets, or sixteen 1-key buckets, and Bucket memory (Buckt_mem) consists of an array of bucket groups. Thus, a bucket group always contains 16 keys. As each key is 128-bit wide, the width of the bucket group is 2048 bits. Bucket memory is an array of bucket groups of different capacities. As shown in Figure 3, a bucket descriptor contains a bucket pointer consisting of a 16-bit offset (Offset) into bucket memory and a 4-bit bucket number (Bucket_no (3:0)). The interpretation of Bucket_no depends on the type of bucket. For a 16-key bucket, Bucket_no (3:0) has no significance since there is only one bucket present in the bucket group. For a 4-key bucket, Bucket_no (3:2) identifies one out of four buckets present in the bucket group. For an 8-key bucket, Bucket_no (3) identifies one of two buckets. For a 1-key bucket, all four bits of Bucket_no (3:0) identify a bucket number, since there are sixteen buckets and there is only one key in each bucket. Furthermore, 128-bit data corresponding to the keys is also stored in BRAM using a data structure similar to bucket memory (refer to Figures 4 and 5). The BRAM, which stores the data corresponding to the keys, is called data memory. The index returned by the comparator mentioned above is used to access the data corresponding to the search key from data memory. The number of 1-key, 4-key, 8-key, and 16-key buckets is software configurable.

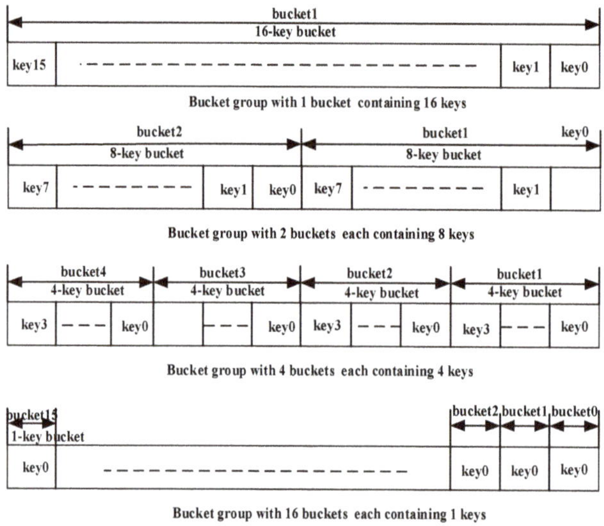

Figure 4. Format of buckets of different capacities and bucket groups.

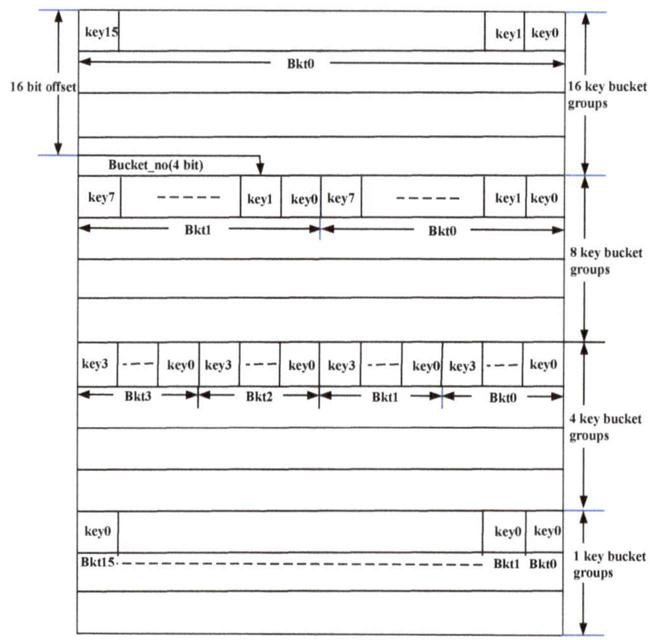

Figure 5. Key Bucket Memory with Bucket Groups.

4.1.1. Architecture of KVS_Bram

The block diagram of KVS_bram is shown in Figure 6. KVS_bram receives a stream of fixed-length keys (128 bits), Command_code (2 bits), Command_ID (2 bits), and data (128 bits) on cmd_ID_Key_data_Interface and provides key, lookup data, command, and error signals on Results_Interface, as shown in Figure 6. The error signals indicate the success or failure of the command issued on cmd_ID_Key_data_Interface. Different sub-blocks of KVS_bram are described below:

1. Command_key_data_ID_FIFO: This FIFO stores a 128-bit key, 128-bit data, and a 2-bit command received on the cmd_ID_key_data_interface. It also stores the 2-bit Command_ID.
2. Hash_val_calculator (Hash Value Calculator): The 128-bit key and command are read by Hash_val_calculator, which computes a 16-bit checksum on the 128-bit key. It uses a pipelined architecture and computes a 16-bit checksum in 4 clocks. The output of this module is a 16-bit checksum, the value of the key, Command_code, and Command_ID.
3. Operations_decode_exec_sm (Operations Decode Execute State Machine): This block decodes 2-bit Command_code and performs all the operations required to execute the search, insert, and delete commands. In the case of insert and delete commands, new 1-key, 4-key, 8-key, and 16zkey buckets are assigned, and one that is freed is put back in the free pool. In case a bucket of particular capacity is full and one more key hashes into it, a higher-capacity bucket is obtained from the free pool, and this bucket is returned to the free pool. Keys from the freed bucket are transferred to that obtained from the free pool. In the case of a search command, the 16-bit hash value is used as an index in the hash table, which contains bucket descriptors as shown in Figure 3. If the bucket descriptor is valid, 16-bit offset and 4-bit bucket_no are used to load keys from the selected bucket. The keys are loaded into a pipelined comparator, which compares the keys in parallel with the input key and returns an index of the matching key. This index is used to read the data corresponding to the key from data memory.
4. Mux and Arbiter—This block multiplexes commands received from the AXI-lite interface and cmd_key_data_interface, as shown in Figure 6.
5. Registers_&_Cmd_Storage—This block stores the insert and delete commands received from the AXI-lite interface through software and returns the status of command execution to software.
6. Cmd_key_data_response_FIFO—This FIFO stores the data returned for the search command and Command_code along with the key and Command_ID. It also stores the command success/error code.

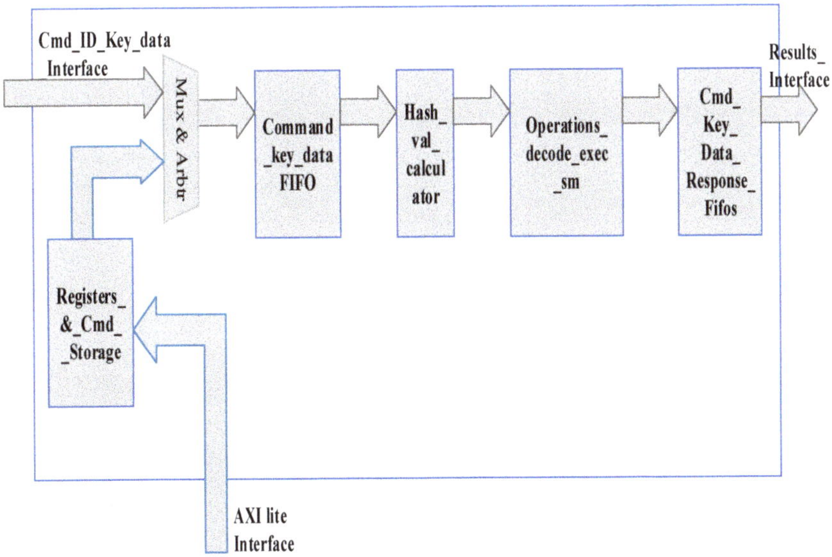

Figure 6. Block Diagram of KVS_bram.

The algorithmic steps of insert and delete operations are described below.

4.1.2. Insert Key Algorithm

1. Read the incoming data and key.
2. Compute the hash of the key.
3. From the hash value, check whether the allocated bucket type is 1, 4, 8, or 16 key type.
4. If (bucket type is type 1) then:
    a. If (a new bucket of type 4 is available) then:
        i. Copy the keys from the type 1 bucket to the type 4 bucket and add the incoming key to this. Return type 1 bucket to free pool.
    b. Otherwise, if (a new bucket of type 8 is available) then:
        i. Copy the keys from the type 1 bucket to the type 8 bucket and add the incoming key to this. Return type 1 bucket to free pool.
    c. Otherwise, if (a new bucket of type 16 is available) then:
        i. Copy the keys from the type 1 bucket to the type 16 bucket and add the incoming key to this. Return type 1 bucket to free pool.
    d. Otherwise, store the keys in HBM.
    e. End if.
5. If (bucket type is type 4) then:
    a. If (key count in the bucket < 4) then:
        i. add the key to type 4 bucket.
    b. If (a new bucket of type 8 is available) then:
        i. Copy all the keys from the type 4 bucket to the type 8 bucket and add the incoming key to this. Return type 4 bucket to the free pool.
    c. If (a new bucket of type 16 is available) then:
        i. Copy all the keys from the type 4 bucket to the type 16 bucket and add the incoming key to this. Return type 4 bucket to the free pool.
    d. Otherwise, store the keys in HBM.
    e. End if.
6. If (bucket type is type 8) then:
    a. If (key count in the bucket < 8) then:
        i. Add the key to type 8 bucket.
    b. If (a new bucket of type 16 is available) then:
        i. Copy all the keys from the type 8 bucket to the type 16 bucket and add the incoming key to this. Return type 8 bucket to the free pool.
    c. Else:
        i. Store the keys in HBM.
    d. End if.
7. If (bucket type is type 16) then:
    a. If (key count in the bucket < 16) then:
        i. add the key to type 16 bucket.
    b. Else:
        i. Copy all the keys from type 16 to HBM and add the incoming key to it. Return type 16 bucket to the free pool.
    c. End if.
8. Else allocate new 1-key, 4-key, 8-key, or 16-key bucket.
9. End.

4.1.3. Delete Key Algorithm

1. Read the incoming key and data.
2. Compute the hash of the key.
3. From the hash value, check whether the allocated bucket type is 1, 4, 8, or 16 key type.
4. If the bucket type is 1:
    i. Delete the key from the bucket.
    ii. Return the bucket to the free pool.
5. If the bucket type is type 4:
    a. If the key count is greater than 1:
        i. Then delete the key from the bucket.
    b. Else if the key count is equal to 1:
        i. Then delete the key and return the bucket to the free pool.
    c. End if.
6. Else if the bucket type is type 8:
    a. If the key count is greater than 5:
        i. Then delete the key from the bucket.
    b. Else if the key count is less than 5 and greater than 1:
        i. Then if a type 4 bucket is available:
            (1) Then copy the keys from the type 8 bucket to the type 4 bucket.
            (2) Return the type 8 bucket to the free pool.
    c. Else if the key count is 1:
        i. If a type 1 bucket is available:
            (1) Then move the key from type 4 to type 1 bucket.
            (2) Return the type 8 bucket to the free pool.
    d. End if.
7. Else if the bucket type is type 16:
    a. If the key count is greater than 9:
        i. Then delete the key.
    b. Else if the key count is greater than 4 and less than 9:
        i. Then if a bucket of type 8 is available:
            (1) Copy the keys from type 16 to type 8 bucket.
            (2) Return the type 16 bucket to the free pool.
    c. Else if the key count is greater than 1 and less than 5:
        i. Then if a bucket of type 4 is available:
            (1) Copy the keys from type 16 to type 4 bucket.
            (2) Return the type 16 bucket to the free pool.
    d. Else if the key count is equal to 1:
        i. Then if a bucket of type 1 is available:
            (1) Copy the key from the type 16 bucket to the type 1 bucket.
            (2) Return the type 16 bucket to the free pool.
8. If the key count is zero, then return the bucket to the free pool.
9. End if.
10. End.

4.2. KVS_hbm

KVS_hbm handles the keys stored in on-chip HBM memory. It uses a 5-level B-Tree for implementing insert, delete, and search operations. Insert and delete operations are implemented in software, while search operations are implemented completely in hardware. The root node of B-Tree and the first two levels are contained in BRAM, while the last two levels are implemented in HBM. The last two levels contain a very large number of nodes and require more storage, so they are located in HBM. The structure of each node of the B-Tree is shown in Figure 7. The node contains an 8-bit count of the number of keys in the node (the maximum number of keys in a node is 33). This is followed by an array of 33 keys (key0–key32) and corresponding data pointers to 33 keys (data_ptr0-data_ptr32). The node also contains 34 pointers to the nodes in the next level (ptr0-ptr33). The hit count denotes the number of times the particular key is hit during a search operation. There are 33 hit counts, Hit_count0—Hit_count32, corresponding to 33 keys. This is used to transfer keys with a large number of hits to BRAM to increase the overall throughput of KVS as explained later. The root node of the B-Tree is located in a register inside FPGA, and the first two levels are implemented in BRAM while the remaining two tree levels are located in HBM. The total number of nodes in the tree is given by the equation:

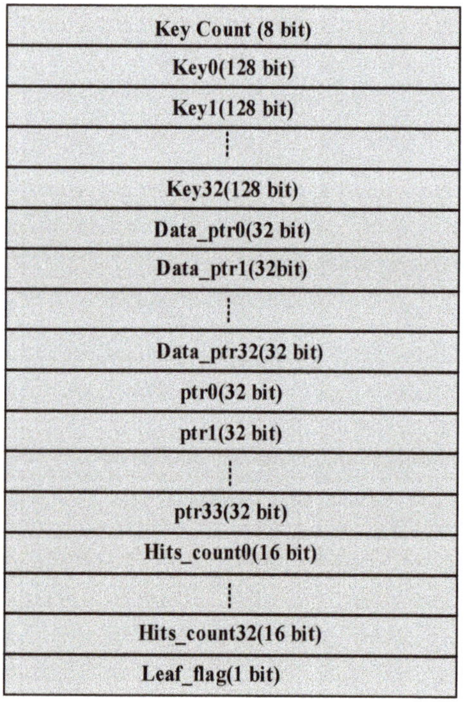

Figure 7. Structure of a B-Tree Node.

The total number of nodes = $33 + 33 \times 34 + 33 \times 34^2 + 33 \times 34^3 + 33 \times 34^4 = 45{,}435{,}423$, which is greater than 40 million nodes as per the requirement.

Architecture of KVS_hbm

A detailed block diagram of KVS_hbm is shown in Figure 8. KVS_hbm consists of the following sub-blocks:

1. Lookup Engines0–3: The four lookup engines work in parallel, and each engine interfaces to 2GBytes of HBM, which contains the key database. The key database is replicated across four banks of HBM connected to four lookup engines by software while executing insert commands. Each lookup engine receives the search command, key, and Command_ID of the command from the Command_assign block. It carries out the search command using a B-Tree. The root node and first two levels of the

B-Tree are implemented in FPGA logic while the nodes in the remaining two levels are stored in HBM. The B-Tree nodes contain 33 keys, which are arranged in ascending order. If a key is found in KVS_bram, the Cmd_ID_alloc_&_reseq block issues an abort command with the ID of the search command, which was previously allocated by the Cmd_ID_alloc_&_reseq block, over Abort_IDs_Inteface. The command_assign block then issues the abort command to the respective lookup engine to which the command is assigned, and the search command is aborted. Lookup engines use highly pipelined 128-bit.

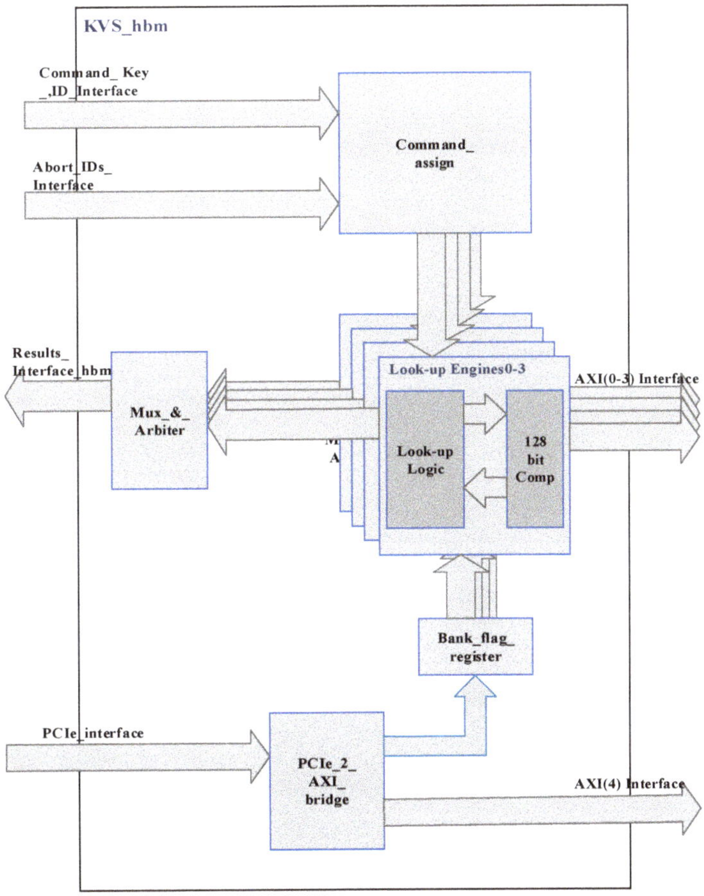

Figure 8. Detailed Block Diagram of KVS_hbm.

A comparator is used to compare the input key with a maximum of 33 keys present in the node of the B-Tree. The comparator returns an index and two flags—equal_flag and right_flag—as a result of the comparison. If the input key is equal to 1 of the 33 keys, the comparator returns the index of the matching key, and equal_flag is set. The index is used to retrieve the pointer data, and using the pointer, data corresponding to the key are read from HBM. If the input key falls between the two keys key(n-1) and key(n), index n-1 is returned and right_flag is set. If the key is less than key(0) in the node, an index of 0 is returned and right_flag is reset. If the input key is greater than key(32), an index of 32 is returned and right_flag is set. Therefore, the comparator returns the index of keys between which an input key lies or the extreme left or extreme right indices. The index is used to retrieve the pointer to the B-Tree node in the next level of the B-Tree. The search operation continues until the matching key is found or the leaf level of the B-Tree is reached as indicated by Leaf_flag (refer to Figure 7). Lookup engines return the lookup result with a success/error status over results_interface_hbm, which is supplied to the Cmd_ID_alloc_&_reseq block.

2. Comamnd_assign: The Command_assign block receives the search command, input key, and Command_ID over Command_Key_ID_interface. It allocates the command to one of the lookup engines that are free. When a command is assigned to the lookup engine, a busy flag is internally set in the engine, which is reset by the respective engine on completion of the command. The command can be completed successfully or with an error status if the key is not found or an abort command is issued. The Command_assign block also keeps track of which Command_ID is allocated to which Lookup Engine and accordingly issues an abort command to the respective engine on the Abort_IDs_Interface.
3. PCIe_2_AXI_bridge: PCIe to AXI bridge converts the PCIe protocol to the AXI protocol and is used for issuing insert commands using software to KVS_hbm. PCIe_2_AXI_bridge is also used to access Bank_flag_register, which maintains bank_flag status. Bank_flag indicates which of the set of 8Gbytes HBM (hbm_set1 or hbm_set2) is active at any time, as explained earlier.
4. Bank_flag_register: This register contains the bank_flag, which indicates whether hbm_set1 or hbm_set2 is active. When hbm_set1 is active, bank_flag is 1 and search operations are carried out on hbm_set1 while insert commands are carried out on hbm_set2. These insert operations are typically performed in bulk. After insert operations are completed on hbm_set2, software resets the bank_flag so that search operations are carried out from hbm_set2 and hbm_set1 is used for carrying out insert operations in bulk. While insert operations are being carried out in hbm_set2, search operations are carried out in parallel on hbm_set1 and vice versa.

4.3. BRAM/HBM Usage for KVS_Top

Each node of the B-Tree requires 864 Bytes of memory storage. Therefore, KVS_hbm requires 3.93 Mbytes BRAM for four lookup engines to store the first three levels of the B-Tree. It requires 1.852 GBytes of HBM for storing B-Tree nodes and 128-bit data corresponding to the keys. KVS_bram requires 30 Mbytes of UltraRAM and 4 Mbytes of BRAM for storing keys, data, and the hash table.

5. Experimental Results

KVS_bram and KVS_hbm were separately synthesized in Xilinx Virtex Ultrascale + VU47P FPGA. Synthesis was performed with Vivado v2021.1. The resource utilization of KVS_bram and KVS_hbm is shown in Tables 1 and 2, respectively.

Table 1. Resource Utilization of KVS_bram.

Flip_Flops/Total/Utilization	LUTs/Total/Utilization	UltraRAM KB/Total/Utilization
39,192/2,607,360 = 1.503%	52,037/1,303,680 = 3.99%	460/960 = 47.92%

Table 2. Resource Utilization of KVS_hbm.

Flip_Flops/Total/Utilization	LUTs/Total/Utilization	BRAM (MB) Total/Utilization
90,329/2,607,360 = 3.461%	146,221/1,303,680 = 11.21%	4.231/8.86 = 47.75%

5.1. Simulation of KVS_Top

Simulation of KVS_Top was carried out using Cadence Xcelium Single Core simulator version 22.0. The architecture of the test bench used for generating search commands is shown in Figure 9. The DUT (Design Under Test) consists of the Verilog model of KVS_Top and models of AXI-Interconnect and HBM. The test bench consists of the following components:

1. CMD/Key_Generator—Generates a 36-bit index randomly. This index is used to look up a table of 128-bit keys stored in the test bench at initialization. The block also generates search commands and sends the key and command to the DUT. The time at which the key and command are sent to DUT is stored in Timestamp_FIFO.
2. Timestamp_FIFO—Stores the time at which search commands are delivered to KVS. The Timestamp stored in FIFO is used to find the latency of the search command execution.
3. Response_Analyzer—Stores the keys and corresponding data in an Associative array which is initialized with values of 128-bit keys and corresponding 128-bit data before the simulation starts. When the DUT sends a response consisting of a 128-bit key and corresponding data, the key is used to look up an associative array, and the value of corresponding the 128-bit data (expected data) is retrieved from the array. This value is compared with the 128-bit data received in the response from DUT to determine the data integrity. Furthermore, the Response Analyzer notes the time at which the response is received and reads the time at which the corresponding command was delivered to DUT from Timestamp FIFO. The difference between the two timestamps gives the latency of the search command execution. Moreover, the DUT response contains a flag showing whether the response is received from KVS_bram or KVS_hbm. In case a response is received from KVS_hbm, the level at which the key is found in the B-Tree is also recorded. These data are used to find the latency of the search command execution when the corresponding data are found at different levels in the B-Tree.
4. Statistics_Report_gen: Maintains the statistics of delay values for SEARCH commands executed by KVS_bram and KVS_hbm. It calculates the average delay values and generates a report of latencies and the throughput of KVS.
5. RC_Model: This block is the model of the PCIe Root Complex and generates PCIe read and write commands for sending insert commands to KVS_hbm by software. It also sends insert and delete commands to the KVS_bram through the AXI-lite interface present in KVS_bram.

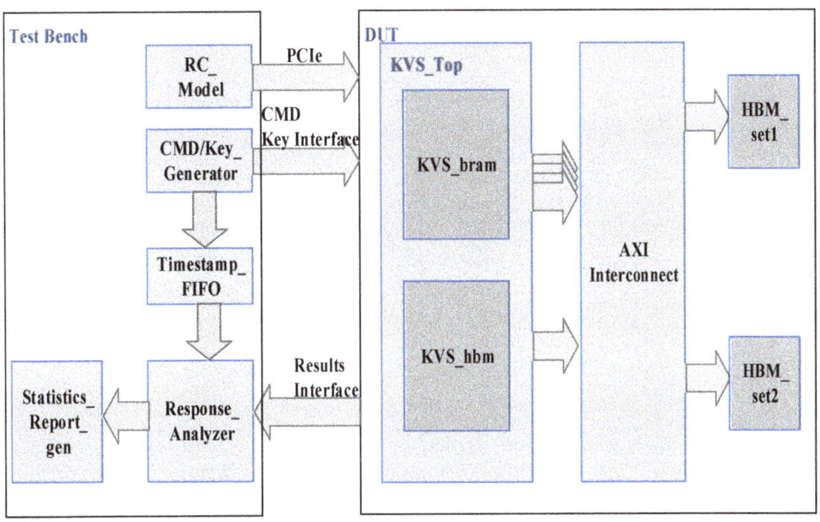

Figure 9. Architecture of Test Bench for Generating Search commands to KVS_Top.

The test bench operation consists of two phases:

1. Initialization Phase—In this phase, BRAM/UltraRAM and HBM memories in the FPGA are initialized with the values of 128-bit keys and corresponding 128-bit data. We have implemented a software function that generates the database of 128-bit keys and data to be inserted into the KVS_bram and KVS_hbm. The function generates the B-Tree nodes in the format shown in Figure 7. It also performs balancing of the B-Tree. The function generates two files for initializing BRAM/UltraRAM and HBM memory.
2. Run Phase—In the Run phase, the search, insert, and delete commands are delivered to KVS_bram and KVS_hbm.

5.2. Simulation Results

The KVS_Top was synthesized at the frequency of 300 MHz, and the timings for insert, delete, and search command execution for KVS_bram are given in Table 3.

Table 3. Command Execution timings for KVS_bram.

Sr No.	Command	Timing
1	Insert	20
2	Search	12
3	Delete	25

Average command Execution timings for the search command for KVS_hbm when the key match is found at different levels of the B-Tree are given in Table 4.

Table 4. SEARCH Command Execution time for KVS_hbm.

Sr No.	Level	Timing
1	0 (root)	170.6 ns
2	1	210.2 ns
3	2	243.8 ns
4	3	371.2 ns
5	4	634.2 ns

5.3. Performance (Search Commands/s) of KVS_Top

The performance of KVS_bram for search commands is 33 million search operations/second. For KVS_hbm, the execution time for the search command depends upon the level at which the key is found. The worst-case performance of one lookup engine is $10^3/634.2 = 1.58$ million search operations/s, assuming the key is found at level 2 in HBM. Therefore, in the worst case, if the key is found at level 4, the performance of KVS_hbm is 1.58 million search operations/s $*$ 4 = 6.32 million search operations/s since four lookup engines are operating in parallel (the performance of one lookup engine is $10^3/634.2 = 1.58$ million search operations/s) since 634.2 ns is the worst-case lookup time. The performance of KVS_Top will depend on the distribution of input keys in the search command.

6. Conclusions and Future Work

In this paper, we present the design of a Key-value Store for a trading system application. By using a combination of Hashing and B-Tree techniques, we can meet the specifications and performance requirements of the stock exchange. To obtain higher performance for active users, we store the key and data corresponding to those users in BRAM/UltraRAM. The KVS_hbm performance can be increased significantly by pipelining the lookup process for different levels of the B-Tree. In future work, we need to design an algorithm that scans the keys in HBM periodically and shifts the keys with a large number of hits from HBM to BRAM/UltraRAM without affecting the throughput of search operations. The number of the keys stored in KVS_bram can also be increased with the use of higher-capacity FPGAs such as Xilinx Versal series FPGAs.

Author Contributions: Conceptualization, S.R. and S.P.; methodology, S.R.; software, M.B.; validation, S.P.; formal analysis, R.P.; writing—original draft preparation, S.P.; writing—review and editing, R.P.; visualization, S.R.; supervision, R.P. All authors have read and agreed to the published version of the manuscript.

Funding: This research received no external funding.

Conflicts of Interest: The authors declare no conflict of interest.

References

1. Hauck, S.; DeHon, A. *Reconfigurable Computing: The Theory and Practice of FPGA-Based Computation (Systems on Silicon)*, 1st ed.; Morgan Kaufmann: Burlington, MA, USA, 15 November 2007; ISBN 0123705223.
2. Available online: https://www.lseg.com/areas-expertise/technology/capital-markets-technology-services/millennium-exchange (accessed on 22 December 2022).
3. Hsiue, K.D. FPGA-Based Hardware Acceleration for a Key-Value Store Database. Master's Thesis, S.B., EECS, Massachusetts Institute of Technology, Cambridge, MA, USA, 2013.
4. Wiggins, A.; Langston, J. Enhancing the Scalability of Memcached. 2012. Available online: http://software.intel.com/enus/articles/enhancing-the-scalability-of-memcached (accessed on 22 December 2022).
5. Qiu, Y.; Xie, J.; Lv, H.; Yin, W.; Luk, W.-S.; Wang, L.; Yu, B.; Chen, H.; Ge, X.; Liao, Z.; et al. FULL-KV: Flexible and Ultra-Low-Latency In-Memory Key-Value Store System Design on CPU-FPGA. *IEEE Trans. Parallel Distrib. Syst.* **2020**, *31*, 1828–1844. [CrossRef]
6. Deepakumara, J.; Heys, H.M.; Venkatesan, R. FPGA implementation of MD5 hash algorithm. In Proceedings of the Canadian Conference on Electrical and Computer Engineering 2001, Conference Proceedings (Cat. No.01TH8555), Toronto, ON, Canada, 4–7 May 2001; Volume 2, pp. 919–924. [CrossRef]
7. Cho, J.M.; Choi, K. An FPGA implementation of high-throughput key-value store using Bloom filter. In Proceedings of the 2014 International Symposium on VLSI Design, Automation and Test, Hsinchu, Taiwan, 28–30 April 2014; pp. 1–4. [CrossRef]
8. Ren, Y.; Liao, Z.; Shi, X.; Xie, J.; Qiu, Y.; Lv, H.; Yin, W.; Wang, L.; Yu, B.; Chen, H.; et al. A Low-Latency Multi-Version Key-Value Store Using B-Tree on an FPGA-CPU Platform. In Proceedings of the 2019 29th International Conference on Field Programmable Logic and Applications (FPL), Barcelona, Spain, 8 September 2019; pp. 321–325. [CrossRef]
9. Bando, M.; Artan, N.S.; Chao, H.J. Flashlook: 100- gbps hash-tuned route lookup architecture. In Proceedings of the International Conference on High Performance Switching and Routing, HPSR 2009, Paris, France, 22–24 June 2009; pp. 1–8.
10. Sourdis, I.; Pnevmatikatos, D.; Wong, S.; Vassiliadis, S. A reconfigurable perfect-hashing scheme for packet inspection. In Proceedings of the International Conference on Field Programmable Logic and Applications, Tampere, Finland, 24–26 August 2005; pp. 644–647.
11. Istvan, Z.; Alonso, G.; Blott, M.; Vissers, K. A Flexible Hash Tabke Design for 10GBPS Key-Value Stores on FPGAS. In Proceedings of the 2013 23rd International Conference on Field Programmable Logic and Applications, Gothenburg, Sweden, 31 August–4 September 2020.
12. Pagh, R.; Rodler, F. Cuckoo hashing. *J. Algorithms* **2004**, *51*, 122–144. [CrossRef]
13. Kirsch, A.; Mitzenmacher, M.; Wieder, U. More robust hashing: Cuckoo hashing with a stash. *SIAM J. Comput.* **2009**, *39*, 1543–1561. [CrossRef]
14. Liang, W.; Yin, W.; Kang, P.; Wang, L. Memory efficient and high performance key-value store on FPGA using Cuckoo hashing. In Proceedings of the 2016 26th International Conference on Field Programmable Logic and Applications (FPL), Lausanne, Switzerland, 29 August 2016; pp. 1–4. [CrossRef]
15. Boutros, A.; Grady, B.; Abbas, M.; Chow, P. Build fast, trade fast: FPGA-based high-frequency trading using high-level synthesis. In Proceedings of the ReConFigurable Computing and FPGAs (ReConFig), Cancun, Mexico, 4–6 December 2017.
16. Ding, L.; Yin, W.; Wang, L. Ultra-Low Latency and High Throughput Key-Value Store Systems Over Ethernet. In Proceedings of the 2018 IEEE International Symposium on Circuits and Systems (ISCAS), Florence, Italy, 27–30 May 2018; pp. 1–5. [CrossRef]
17. Puranik, S.; Sinha, S.; Barve, M.; Patrikar, R.; Shah, D.; Rodi, S. Key-Value Store using High Level Synthesis Flow for Securities Trading System. In Proceedings of the 2020 International Conference on Computing, Electronics & Communications Engineering (iCCECE), Southend Campus, UK, 17–18 August 2020.

Disclaimer/Publisher's Note: The statements, opinions and data contained in all publications are solely those of the individual author(s) and contributor(s) and not of MDPI and/or the editor(s). MDPI and/or the editor(s) disclaim responsibility for any injury to people or property resulting from any ideas, methods, instructions or products referred to in the content.

electronics MDPI

Article

Acceleration of Trading System Back End with FPGAs Using High-Level Synthesis Flow

Sunil Puranik [1,*], Mahesh Barve [1], Swapnil Rodi [1] and Rajendra Patrikar [2]

1 Tata Consultancy Services, Pune 411057, India
2 Visvesvaraya National Institute of Technology, Nagpur 440012, India
* Correspondence: sunilsavitap@students.vnit.ac.in

Abstract: FPGA technology is widely used in the finance domain. We describe the design of a financial trading system order processing component using FPGAs, implemented with high-level synthesis (HLS) flow. The order processing component is the major contributor to increased delays and low throughput in the current software implementation of trading systems. The objective of FPGA implementation is to reduce the latency of order processing and increase the throughput of trading systems as compared to software implementation. Our design is one of the first attempts to speed up order processing in a trading system using FPGA technology and HLS flow. HLS was used in implementing the design for higher productivity and faster turnaround time. The design shows orders of magnitude of improvement in performance indicating that more complex FPGA systems could be designed using HLS. We obtained more than 2X of an advantage in order processing speed and a reduction in latency with FPGA technology. Moreover, we gained a 4X advantage in terms of productivity using HLS.

Keywords: accelerator architectures; field programmable gate arrays; high-level synthesis; system performance; TCPIP

Citation: Puranik, S.; Barve, M.; Rodi, S.; Patrikar, R. Acceleration of Trading System Back End with FPGAs Using High-Level Synthesis Flow. *Electronics* **2023**, *12*, 520. https://doi.org/10.3390/electronics12030520

Academic Editor: Akash Kumar

Received: 2 December 2022
Revised: 27 December 2022
Accepted: 3 January 2023
Published: 19 January 2023

Copyright: © 2023 by the authors. Licensee MDPI, Basel, Switzerland. This article is an open access article distributed under the terms and conditions of the Creative Commons Attribution (CC BY) license (https://creativecommons.org/licenses/by/4.0/).

1. Introduction

Securities trading systems involve the processing of orders that are generated by end users. These orders are typically placed at a rate of 1 million orders per second and are expected to be processed at very low latencies. Since all the components of a trading system such as order validation, lookups, and order matching are implemented in software in a traditional trading system, the order processing rate is low and the latencies of order processing are high. Furthermore, physical network delays and TCP/IP stack delays add to the software delays, resulting in high latencies and low order processing throughput. So, the idea is to speed up the operation of trading systems by migrating the functionality of trading system components including order validation, order matching, lookups, and TCP/IP stack processing from software to hardware.

1.1. Use of FPGAs for Accelerating Trading Systems

The number of trading systems is not very large (around 60 stock exchanges in the world) [1] and trading systems contain modules that need frequent reconfigurations of their algorithms as well as parameters. For example, business logic in stock exchanges requires frequent changes, such as the addition of multi-leg order commands (a multi-leg options order refers to any trade that involves two or more options that are completed at once). Since volumes are low and functionality requires frequent changes, the use of ASIC technology is not justified for trading systems acceleration, and reconfigurable computing devices such as FPGAs are the best choice for the acceleration of trading systems.

FPGAs [2] are increasingly receiving traction in the field of financial processing where there is a need for frequent changes in business logic and operating parameters such as the load and number of securities to be traded. Added to this, there could be a need for adding newer algorithms to the existing system to make it more intelligent.

The development time taken by classical VHDL/Verilog-based flows is very long and productivity is low [2]. There has been a search for alternate flows which can reduce the development time. High-level synthesis (HLS) [3–6] provides a level of abstraction higher than Verilog and VHDL. HLS can be used to describe algorithms in C/C++ and convert them to digital circuits [3]. Additionally, the productivity gained by HLS is orders of magnitude greater than by traditional methods [7]. HLS is supported through its products by a number of VLSI vendors such as Vivado HLS by Xilinx [8], HLS by Intel Altera [9], and Catapult by Mentor [10]. All these products provide tools for writing code in high-level languages such as C/C++/System C and converting them to Verilog/VHDL.

HLS has been traditionally used for implementing algorithmic workflows making use of C language. HLS finds use in domains such as image processing and high-frequency trading (HFT). Boutros et al. [11] described the usage of HLS for designing an HFT system. Here, we use HLS for speeding up the trading system itself.

As shown in Figure 1 below, the trading system environment consists of users/traders submitting trade requests and a trading system which is located in the stock exchange. While HFT trading provides high-speed processing for users submitting trade requests and sits on the user side, our objective in this paper is to accelerate the trading system itself. This paper uses HLS to migrate the functionality of trading system components which are currently implemented in software, to FPGA hardware. This is performed to reduce latency and increase throughput.

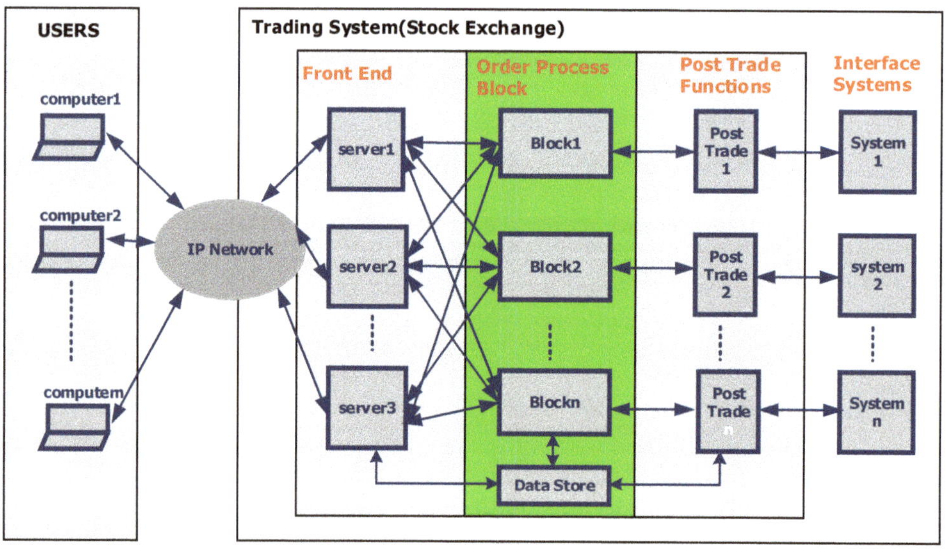

Figure 1. Trading system architecture (the block being accelerated is shown in green color).

1.2. Study Contributions

The main contributions of this paper are as follows:

- To increase the order processing rate of the trading system from 1 million orders/sec (achieved with software implementation) to 2 million orders/sec with FPGA technology.
- To reduce the latency of order processing commands from 1 microsecond (achieved with software implementation) to less than 500 ns. The throughput and latency numbers for software implementation have been taken from a large stock exchange.

- Additionally, as an important feature of our design, to optimize the use of block RAM (BRAM) which is a fast on-chip memory inside the FPGA, by the innovative design of the data structures.
- Through, this design, to also increase the throughput of the UPDATE command by 30–40% using pipelined execution as explained in later sections.

We describe how HLS was used for implementing the three main commands INSERT, UPDATE and DELETE in the trading system back end.

This paper is organized as follows. In Section 2 we describe the related work conducted in this field. Section 3 describes the general architecture of a trading system. In Section 4, we describe the problem statement. Section 5 describes the data structures used in the design and design implementation. Finally, Section 6 describes the performance numbers, followed by Section 7 on the pipelined execution of the UPDATE command, and Section 8 with the Conclusion and Future Work.

2. Related Work

There have been many examples in the literature of FPGAs being used for accelerating financial systems, databases, as well as network protocols. They have found use in high-frequency trading (HFT) [12–14]. These are optimized to achieve the lowest possible latency for interpreting market data feeds. FPGA acceleration for HFT has been described in [15]. FPGA implementation using HLS for HFT has been described in [11]. The study in [16] describes the design of a hardware accelerator to speed up the data filtering, arithmetic, and logical operations of a database. The study in [17] describes the acceleration of a TCP/IP stack with an FPGA. However, after an extensive literature review, we could not find any related previous work that describes the acceleration of a trading system front end and back end with an FPGA.

Trading systems have traditionally existed within the software. Trading system software is very much multi-threaded and is usually found in Linux OS [18,19]. The software makes use of hardware features such as pipelining and multicore technologies. There have been very few instances of the use of FPGAs for a complete trading system back end. A very well-known example of the deployment of FPGAs is in the London Stock Exchange [20]. The system promises extensibility and reconfigurability.

There is very little literature regarding the internal architecture of securities trading systems. This is because these details are mainly proprietary in nature. Moreover, there are very few companies in this field, and revealing the internal architecture could dent their competitive advantage. Hence, the architecture details are not published by the trading system developer firms. Due to this, we were not able to compare the performance of an FPGA-based system to other systems. However, we compared the performance of our system to existing software-based systems.

Our paper describes a trading system accelerator design. FPGAs provide a lot of flexibility that can be exploited by programmers and hardware designers to build accelerators. In data analytics, FPGAs are suited for repetitive tasks. They have been incorporated into platforms such as Microsoft's Project Brainwave [21], the Swarm64 Database Accelerator [22], Postgres [23], the Xilinx Alveo Data Center Accelerator [24], and Ryft [25]. Key–value stores [26] have also been accelerated using FPGAs. Also, FPGAs have become a good option for accelerating databases [27].

3. Trading System Architecture at a High Level

The architecture of the system is depicted in Figure 1. It consists of a number of users/traders connected to the trading system using an IP-based network. These traders are outside the premise of the trading system. The trading system itself consists of three components:

1. Front end
    a. Connects the traders to an IP network.
    b. Accepts orders from users.

   c. Performs validations.
2. Back end (order processing block)
   a. Performs the function of order matching, i.e., matching sell orders with buy orders and vice versa. It maintains the database of the sell and buy orders received from users and executes commands to perform order matching.
   b. Connects to the front end via an Ethernet IP network.
3. Post trade block Once trading is complete, the post trade block performs functions such as:
   a. Journaling;
   b. Recordkeeping;
   c. Sending a response back to a user on an IP network.

As stated above, our objectives are

1. To reduce the latency of order processing;
2. To increase the throughput by implementing order processing functions in FPGA hardware.

Both the front end and order processing block (back end) functions, shown in Figure 1, are implemented in the FPGA using a PCIe-based front end processor board and back end processor board. This architecture is shown in Figure 2.

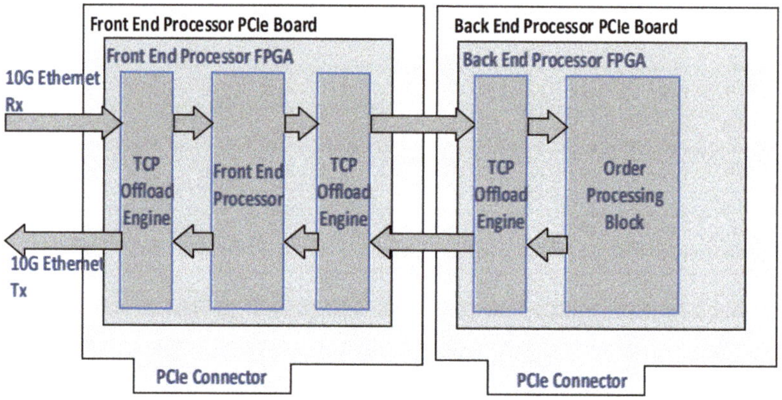

Figure 2. Architecture of the trading system implemented in an FPGA.

The users connect to the front end processor board on the 10G Ethernet network and submit trade requests. The front end processor board uses a TCP offload engine (TOE) block to perform TCP/IP processing in hardware to reduce the network latency. It contains the front end processor FPGA. A block diagram of the front end processor FPGA is shown in Figure 3. It contains a TOE which interfaces to users, validations logic, lookups logic, a connections management block, and a TOE for interface to the back end processor board. Validations logic checks the ranges of different fields in the order request submitted by users and verifies that these fields have valid values. Lookups logic performs many lookups to verify that the data in the different fields in the order request matches the master data. The validations and lookups are performed in parallel by FPGA logic to reduce latency. Connections management logic maintains a table of TCP connection IDs against the user IDs and ensures the response from the back end processor board to a user request is sent on the same TCP connection ID on which the order was received. The second TOE performs the function of interfacing with the back end processor FPGA board.

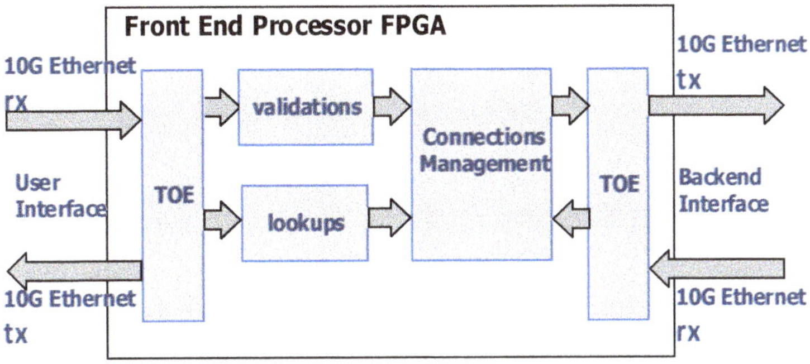

Figure 3. Block diagram of the front end processor FPGA.

The back end processor PCIe board connects to the front end processor board on 10G Ethernet and performs order matching functions in hardware to reduce processing latencies. The block diagram of the back end processor board is shown in Figure 4. It consists of a TOE for interface to the front end processor PCIe board and order processing block, which in turn consists of business logic and a command execute block. The business logic matches the sell orders against the buy orders. For example, if there is a buy order for a particular security at a given price and if it matches the sell order with a lesser price for the same security, the trade will be executed. If there is no matching sell order, then the buy order will be inserted into the database. When trade happens, orders are either deleted from the order database or updated in the order database based on order matching quantities. Command execute logic maintains a linked list of orders and performs a deletion, insertion, or update of orders in the order database as described in later sections.

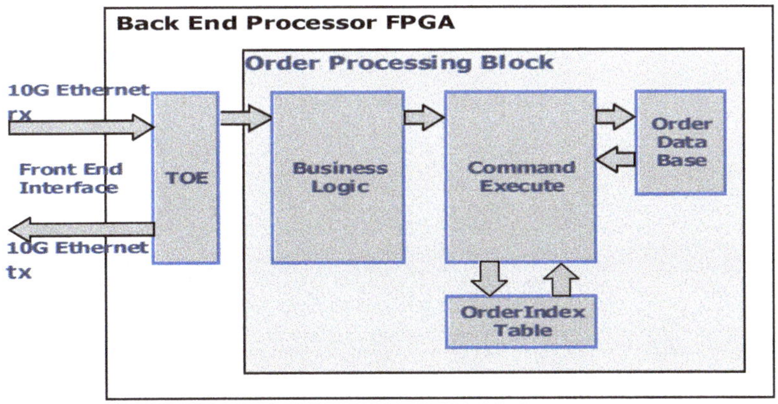

Figure 4. Block diagram of the back end processor FPGA.

To reduce network latency, both boards perform TCP/IP processing in hardware using the TOE block. In our implementation, we focus on the order processing block, which is the most critical block in the system in terms of latency experienced for trade requests submitted by users. Order processing involves the implementation of a number of commands out of which INSERT, DELETE, and UPDATE order are the most frequent ones and occur 95% of the time. So, only these three commands are considered for modeling the order processing block:

1. The INSERT order command, when submitted to the trading system back end, requires the incoming order to be placed in the back end order database by appropriately manipulating the internal data structures.

2. The UPDATE order command requires the system to change the price of the already placed order to a new value. Thus, the order data structure stored in the memory is manipulated to indicate the new price to be used. This is the most commonly executed command occurring more frequently than INSERT and DELETE. Hence, it is necessary that the data structures and modules are designed such that the latency of this command is minimized.
3. The DELETE order command requires that the order placed using the INSERT order command is removed from the order database and the order is not manipulated any further.

4. Problem Statement

As described in Section 2, the trading system consists of a 10G network, front end, and back end blocks. The total order processing latency consists of three components:

1. Network Latency—This consists of TCP/IP processing delays and wire delays. The TCP/IP processing latency is reduced by implementing TCP/IP processing using a TOE block in hardware as mentioned in Section 3. This reduces latency from 3–4 microseconds (required by the TCP/IP stack implemented in software) to around 100 ns.
2. Front End latency—This consists of delays involved in validations and lookups which are performed by the front end. This is reduced by performing validations and lookups in parallel in FPGA logic.
3. Back End (Order Processing) Latency—This delay is the time required for processing the INSERT, DELETE, and UPDATE commands as described above. This paper describes the implementation of an order processing block in an FPGA using HLS to reduce this latency component.

The INSERT, DELETE, and UPDATE orders form the major chunk of the commands executed in the trading system. Any acceleration of the trading system would require the acceleration of these three commands. Thus, the problem at hand is to increase the throughput of the system and reduce the latency of these transactions. To tackle this problem, newer data structures and algorithms are needed. The constraints for implementing this logic in an FPGA are the on-chip memory (BRAM) and FPGA resources.

5. Data Structures

The implementation of the trading system involves the use of the following data structures:

(A) Order Database

The order database stores all the fields and attributes of the order which are placed by the end users. The order structure has all the details needed for processing a transaction. Referring to Figure 5, orders are stored in the order database, which is an array of around one million order structures, stored in Static RAM (a special category of RAM) for fast access. Typical fields in the order structure are the price, time stamp, volume, security identifier, buy/sell flag, OrderID, etc. The offset of the order in the order database is called the OrderIndex and order indexes for all the orders are stored in the order index table shown in Figure 5.

Each order is identified by a unique 32-bit OrderID and this OrderID is used to address the order index table. For example, if orders Order0, Order1, ... OrderK stored at offsets 0, 1, ... k, respectively, as shown in Figure 5, have OrderIDs m0, m1, ... mk, then integer 0 is stored at address m0, integer 1 is stored at address m1 and integer K is stored at address mk in the order index table. This way, using OrderID in an incoming order, the index of the order can be obtained from an order index table lookup and the OrderIndex can be used to locate the order in order database. The order index table is stored in DRAM as OrderID is 32-bit, which requires 4GB of storage.

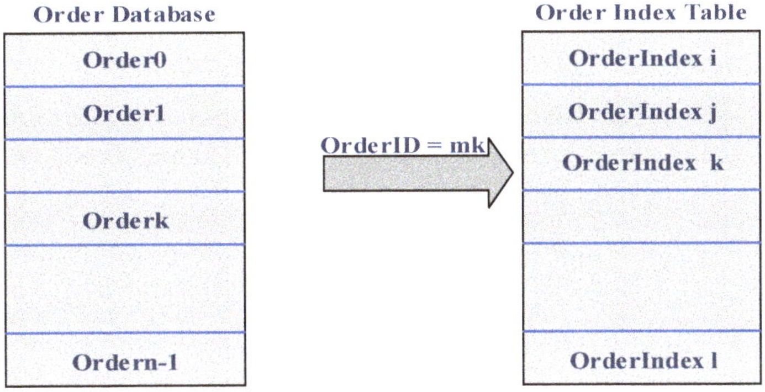

Figure 5. Order database and order index table.

(B) Security Pointers Table

Order nodes are used to store important and frequently accessed information about an order (for example, the price, OrderID, buy/sell flag, etc.). There is one order node corresponding to every order in the order database. For faster access, order nodes are stored in BRAM at the same offset as the corresponding order in the order database. (This offset is the same as the OrderIndex in the order index table). Since order nodes contain only the frequently accessed information about the order, the use of block RAM (BRAM) is optimized. Each security is identified by a unique TokenID. The head pointer to each security linked list is stored in the securities pointer table as shown in Figure 6, at the offset equal to the TokenID. For each security, order nodes for a particular price are stored in a vertical linked list, as shown in Figure 6. They are sorted according to the time stamps. For a given security, there is a vertically linked list of order nodes for each price point. The price point information is stored in a dummy order node and these dummy order nodes are arranged as a horizontally linked list. The dummy order nodes or price points are arranged in the decreasing order of prices for buy orders and in the increasing order of prices for sell orders. Pointers or offsets to the order nodes and dummy order nodes are stored in a free pool, which is accessed as first-in, first-out (FIFO). FIFO stores the offsets of order nodes and dummy nodes. A pointer to the new order node is obtained during an INSERT command execution from the free pool and the pointer is returned to the free pool during the execution of the DELETE command.

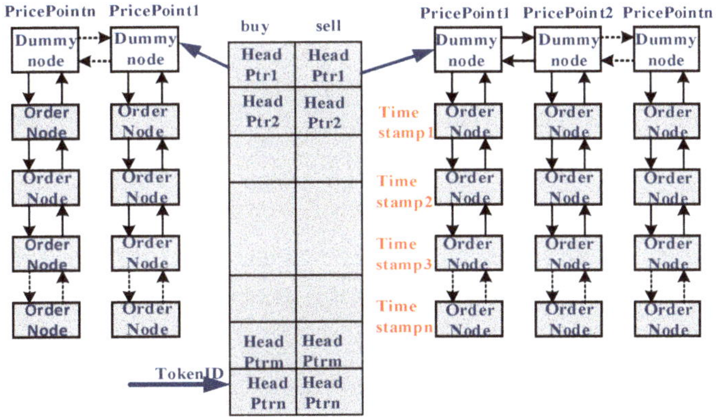

Figure 6. Securities pointer table.

6. HLS Implementation of the Order Processing Block

A block diagram of the order processing block which executes the INSERT, UPDATE, and DELETE commands is shown in Figure 7 below. The order processing block consists of the following components:

1. Command Queue—The command queue is used to store the commands delivered on the command interface.
2. Command Decode Logic—This block reads the commands from the command queue and decodes them. Based on the command code, it calls the different functions in the command execute block to execute the command.
3. Command Execute Block—This block contains all the subfunctions required to execute the INSERT, DELETE, and UPDATE commands as explained later.
4. Free Pool FIFO—The Free Pool FIFO stores the pointers to free order nodes and dummy order nodes.
5. Dummy Order Node Array—The dummy order node array is used to store the linked list of dummy order nodes which contain the price information of buy and sell orders. They are stored in BRAM for faster access.
6. Order Node Array—The order node array is used to store the linked list of order nodes which contain the frequently accessed information about the orders. These are stored in BRAM for faster access.
7. Order Index Table—As explained earlier, the order index table is used to store OrderIndex information and is accessed by OrderID. This table is implemented in DRAM.
8. Command Status Queue—This contains the status of the commands that were delivered on the command interface.
9. Order Database—The order database contains the orders placed by users of the trading system. It is stored in SRAM for faster access.

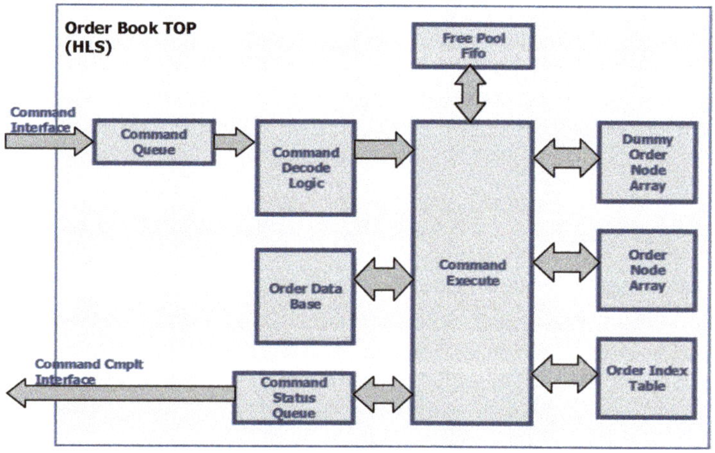

Figure 7. Block diagram of the order processing block implemented in HLS.

The INSERT, UPDATE, and DELETE commands delivered over the command interface are stored in the command queue. The command decode logic reads the commands from the queue, decodes the commands, and calls the command execute logic functions to execute the required command. Command execute logic implements HLS functions to read the order node from the free pool FIFO, return the order node to the free pool, insert the order node into the order node array, remove the order node from the order node array, read the OrderIndex from the order index table, and manipulate the pointers for inserting and deleting the order nodes in the order node arrays. The dummy order node array and order node array are implemented as doubly linked lists as shown in Figure 6. As the order index table is stored in DRAM and the order database is stored in SRAM, which are both off-chip memories, the AXI master interface is used for accessing these data structures. We used the pragma HLS interface m_axi port = ord_ind_arr for implementing the AXI master interface. We also used the pragma HLS interface bram port = ord_nd_arr and pragma HLS interface bram port = dmy_nd_arr for implementing BRAM interfaces for order node and dummy order node arrays, respectively. To implement pipelined operations, pragma HLS pipeline II= n was used. The pipeline pragma was also used to pipeline the loops and obtain a higher frequency operation. Pragma HLS latency = max_value was used for constraining latency values.

Table 1 below shows the HLS pragmas used in the code in tabular form.

Table 1. Details of the pragmas used in the HLS code for the order book top.

Sr. No.	Block Name	Pragma	Value
1	Order_book_top	HLS Interface	m_axi port = ord_ind_arr
2	Order_book_top	HLS Interface	m_axi port = ord_bk_arr
3	Order_book_top	HLS Interface	Bram port = ord_nd_arr
4	Order_book_top	HLS Interface	Bram port = dmy_nd_arr
5	Command Execute	HLS Pipeline	II = 1
6	HLS_top	HLS Latency	Max 200
7	Command Queue	HLS Stream	Depth = 8
8	Command Status Queue	HLS Stream	Depth = 8

The steps involved in executing the INSERT, UPDATE, and DELETE commands by the order processing block are described below.

The order processing block (back end logic) decodes the orders received from the front end and takes the following steps during the execution of each of the INSERT, UPDATE, and DELETE order commands:

A. INSERT Order

1. Get the pointer to the new order node (offset of the order node) from the free pool FIFO.
2. Store the order structure in SRAM at the same offset (offset obtained in step 1) as the order node.
3. Make the entry in the order index table in DRAM. Write the offset of the order in DRAM (which is the same as the offset of the order node in BRAM) in the order index table using the OrderID as the address. (The OrderID is received as a part of the order request). Set a flag to indicate that the content of the location is valid.
4. Traverse the dummy order nodes linked list to find the location where the order node corresponding to the new order can be placed based on the price field in the order.

B. UPDATE Order

1. Using the OrderID field in the request as an address, read the OrderIndex (offset of the order in SRAM) from the order index table stored in DRAM memory.
2. Using the OrderIndex, the order node corresponding to the order is accessed.
3. This order node is moved to the new price point position and added at the end of the vertical linked list of order nodes and deleted from the current price position in the vertical linked list under the dummy node corresponding to the old price position. If there is only one order node under the dummy node corresponding to the old price position, the dummy node is deleted and returned back to the free pool FIFO.

4. If the dummy node corresponding to the new price position does not exist, it is obtained from the free pool and added to the horizontal linked list, and the order node is added at the head of the vertical linked list under the newly added dummy node.

C. DELETE Order

1. Using the OrderID field in the request as an address, read the OrderIndex from the order index table stored in DRAM. Reset the flag in the DRAM location indicating that the location content (OrderIndex) is no longer valid.
2. Locate the order node using the OrderIndex. The order node is at the same offset in block RAM (BRAM) as the order structure in SRAM and this offset is equal to the OrderIndex. Remove the order node from the linked list by manipulating the pointers in the time vertical linked list. If it was the only node under the dummy node, then remove the corresponding dummy node as well. Return the order node back to the free pool FIFO.

The algorithmic steps for executing the INSERT, UPDATE, and DELETE commands are shown in Figure 8 below:

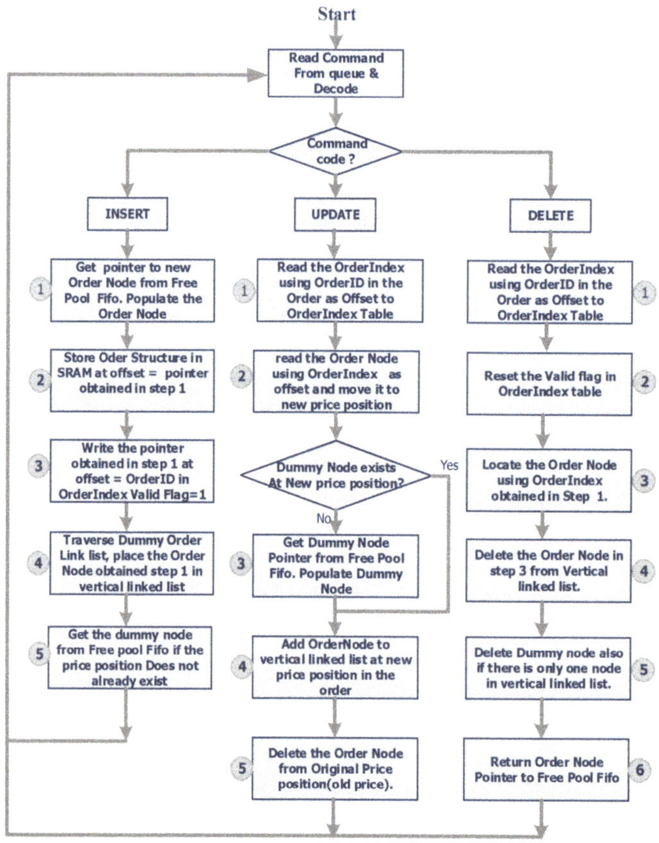

Figure 8. Algorithms for the execution of the INSERT, UPDATE, and DELETE commands.

7. Performance Numbers

To obtain the performance numbers, we implemented the order processing block in HLS, Verilog, and software. The setup consisted of a Vivado HLS IDE and QuestaSim simulator for SystemVerilog. HLS code was run, first in the Vivado HLS IDE environment to confirm logical correctness. Co-simulation was conducted to understand whether the generated Verilog also worked correctly. The HLS code was synthesized on a Xilinx Ultrascale+ FPGA board (Virtex UltraScale+ VCU118-ES1 Evaluation Platform with xcvu9p-flga2104-2L FPGA). It was synthesized with a clock cycle of 3 ns. Latency numbers for overall processing (front end + order processing block) were computed for the system using C-RTL co-simulation in Vivado and the use of SystemVerilog simulations under QuestaSim. The software implementation of the front end and the order processing block consisted of C code run on a fault tolerant machine with a Red Hat Linux 7, 16-core CPU (Intel (R) Xenon(R) CPU ES-2667 V3 @ 3.20 GHz), and 256GB of RAM. The data structures used in the software were different since there was no consideration of the block RAM for software implementation. The software uses hashmap and treemap data structures for the order book. As for the synchronization, single-writer principles were followed to avoid locking contentions in the performance critical path. In a few scenarios, compare and swap low latency locks were used. Dedicated isolated cores were assigned to every process to avoid CPU switching. Due to the sequential nature of software, having more CPU cores did not give significant performance improvement.

The performance of the design was measured based on various parameters, namely, resource utilization, latency, and throughput. Overall, the time required for processing one order was around 500 ns for HLS and Verilog, while it was around 1 microsecond for software. Table 2 gives the details of the resource utilization of the final system after synthesis in the xcvu9p-flga2104-2L FPGA. These details of the resource utilization were made available by the Vivado HLS synthesis tool.

Table 2. Resource utilization.

Flip Flops/ Total/ Utilization	LUTs/ Total/ Utilization	Memory kb/ Total/ Utilization
10629/ 2364480/ 0.45%	18769/ 1182240/ 1.58%	105/ 4320/ 2%

7.1. *Setup for Latency and Performance Measurement*

A block diagram of the test bench and design under test (DUT) for measuring the latency and performance of different commands is shown in Figure 9. The DUT consists of Verilog code of the order processing block implemented in HLS. The test bench consists of the command generator, command queue, command latency mean execution time calculator, and report generator. The command generator generates a random mix of INSERT, UPDATE, and DELETE commands using SystemVerilog constrained random generation and submits the commands to the DUT on the command interface. The weighted random distributions of different commands are generated using the *dist* operator of SystemVerilog random generation. Commands are also queued into the command queue as they are submitted to the DUT along with the time stamp. Commands are executed by the DUT in the order in which they are submitted. The DUT indicates that command execution is complete with the Cmd_cmplt pulse shown in Figure 9. This signal is used to record the command completion time. Commands are retrieved from the command queue in FIFO order, and the command execution time and command latency are calculated by the mean time and latency calculate block, respectively. This block also maintains the count of how many UPDATE, DELETE, and INSERT commands were executed in a particular test case. The report generator prints the report of the latency and command mean execution time for the INSERT, UPDATE, and DELETE commands.

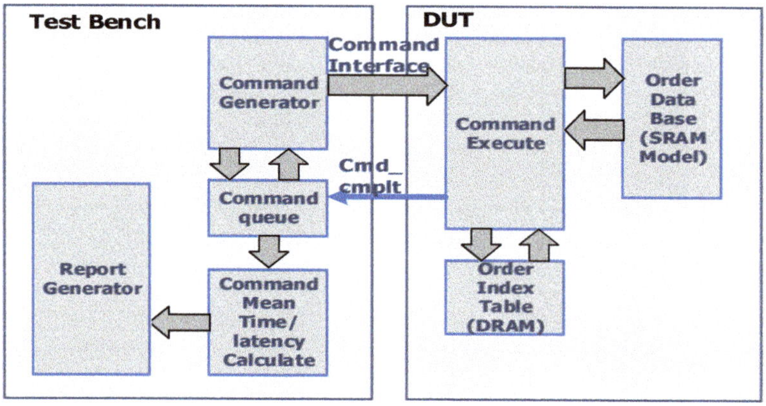

Figure 9. Block diagram of the SystemVerilog test bench and DUT.

7.2. Latency Measurements

The following were our observations with regard to latency for each of the commands:

1. DELETE Order The latency for the DELETE order operation remains constant. It does not change based on the order number or the location of the order in the buy/sell linked list.
2. INSERT Order

The INSERT operation involves traversing the buy/sell dummy order node linked list and placing the incoming order under the appropriate dummy order node. If needed (if the price point does not exist), a new dummy order node may be inserted, and the incoming order placed under this dummy node. Thus, we see that the time for the INSERT order is dependent on the time spent traversing the buy/sell linked list or the number of dummy nodes (hops) that have to be inspected. Thus, latency depends directly on the number of hops.

3. UPDATE Order

The UPDATE operation involves placing one timestamp node in a new location in the buy/sell linked list. The latency is dependent on the number of hops from the current dummy node location to the new dummy node location.

To compute latency, two types of traffic were generated for the system:

a. Sequential traffic that gave a fixed sequence of INSERT, UPDATE, and DELETE commands.
b. Random traffic that gave INSERT, UPDATE, and DELETE commands in some weighted proportion. The proportion was configurable.

These timings have been observed with the QuestaSim SystemVerilog Simulator designed by Mentor Graphics.

7.3. Atomic Transaction Level Latency

From the latency test cases, the following latencies (refer to Table 3) have been observed for the INSERT, UPDATE, and DELETE commands. (These numbers were calculated from sequential traffic tests by giving few INSERT, DELETE, and UPDATE commands).

Table 3. Latencies of various commands.

Command Name	Latency (Clocks)	Comment
INSERT	52	This is for inserting one price point after the initial insertion.
DELETE	45	This timing is constant, irrespective of the number of price points, as expected.
UPDATE	36	This is for the first UPDATE with one hop.

7.4. Formulae for the Expected Latency

For the INSERT/UPDATE commands, the linked list has to be traversed. The number of dummy nodes between the start and the end node is called hops. After running sequential and random traffic tests, we observed the following relationship between the number of hops and corresponding latency for each command:

- INSERT: Clocks for N hops = 50 + 2xN
- UPDATE: Clocks for N hops = 34 + 2xN
- DELETE: Total number of clocks = 45

N in the above formulae is the number of hops. Here, latency is the number of clock cycles with the clock having a period of 3 ns. These latencies were calculated with C-RTL co-simulation and do not include DRAM access latency and DDR controller latency. As expected, the latency of INSERT and UPDATE was proportional to the number of hops while the DELETE latency was constant irrespective of the number of hops.

7.5. Observed Latency under Various Price Depths (Hops)

This study is applicable to UPDATE commands.

Table 4 above shows the latency numbers for 1200 total commands of which the first 300 were inserts and the rest were random where the percentages of INSERT, UPDATE, and DELETE were 10%, 80%, and 10%, respectively.

Table 4. Latencies of various depths for UPDATEs.

Max. Hop	Min. Latency (Clocks)	Max. Latency (Clocks)	Avg. Latency (Clocks)	Std. Dev. Latency (Clocks)
20	70	149	89	10
30	70	159	96	15
40	70	165	100	16

Table 5 below has 1200 total commands of which the first 300 were INSERTs and the rest were random where the weights of the INSERT, UPDATE, and DELETE commands entered were 5%, 90%, and 5%, respectively. From the tables, we can conclude that the minimum time was for the first UPDATE. It can be inferred from the table that irrespective of the distribution, the average latency and maximum latency depend on the number of hops, while the minimum latency remains constant as expected.

Table 5. Latencies of various hops for UPDATEs.

Max. Hop	Min. Latency (Clocks)	Max. Latency (Clocks)	Avg. Latency (Clocks)	Std. Dev. Latency (Clocks)
20	70	151	90	10
30	70	171	97	15
40	70	169	101	17

7.6. Throughput

The throughputs for the INSERT, UPDATE, and DELETE commands were calculated based on the mean time required for the execution of the commands for a given number of hops. The mean time gives the time it takes in ns to execute the command. The throughput of the system was computed under various kinds of loads. The following table, Table 6, depicts the throughput for the INSERT, UPDATE, and DELETE commands. The hops in the table indicate the initial depth of the linked list which vary according to the INSERT and DELETE traffic.

Table 6. Throughput of various commands.

Sr. No.	Command	Hops	Execution Times (ns)	Throughput (commands/sec)
1	INSERT	20	367	2724.79×10^3
2	INSERT	30	361	2770.08×10^3
3	INSERT	40	367	2724.79×10^3
4	UPDATE	20	270	3703.703×10^3
5	UPDATE	30	291	3436.4×10^3
6	UPDATE	40	303	3300.330×10^3
7	DELETE	20	153	6535.9×10^3
8	DELETE	30	162	6172.8×10^3

So, the throughput for the command is
$(10^9)/(\text{mean command execution time})$ commands per second

Note: This calculation was obtained with traffic from 90% UPDATE, 5% INSERT, and 5% DELETE commands.

The software implementation of the INSERT and UPDATE commands takes 1.5 microseconds (0.67 million commands/sec) while DELETE takes 1.2 microseconds (0.84 million commands/sec). It can be seen that with an FPGA, the throughput of all the commands increased to more than 2 million commands/sec (which is the same as orders/sec). Furthermore, the average latency of the command execution was reduced to around 300 ns.

7.7. Productivity

Verilog implementation took 6 months while HLS implementation took 1.2 months with two engineers with experience of 8–10 years.

8. Pipelined Execution of UPDATE Command

The execution of the UPDATE command involves two phases: 1. Fetch—In this phase, the OrderIndex of the order which is stored in the order index table is fetched using the OrderID as the address. Since the order index table is stored in DRAM, the fetching of the OrderIndex takes longer compared to BRAM. 2. Execute Phase—This phase involves moving the order node to a new price position, adding at the end of the vertical linked list of order nodes and deleting the order node from the current price position in a vertical linked list under the dummy node corresponding to the old price position. If the fetch and execute phases are carried out sequentially without any overlap, the execution time of the UPDATE command increases, resulting in less throughput for UPDATE. To address this issue, we modified the UPDATE command execution logic such that there is an overlap between the execution of the fetch and execute phases. This was achieved by executing the fetch and execute phases in a pipelined fashion. The fetch logic looks up the order index table using the OrderID as the address and writes the OrderIndex read from DRAM into a first-in, first-out (FIFO) queue. The execute command reads the OrderIndex from the queue and uses it to locate the order in the order database. After locating the order node, the execute logic moves it to the new position in the horizontally linked list of price nodes. Since the fetching of the OrderID for the next UPDATE command overlaps with the execute phase of the previous UPDATE command, the effective execution time of the UPDATE command is reduced significantly, if the UPDATE commands are received sequentially. Since the percentage of UPDATE command is very high (around 90% of the total commands are UPDATEs), this modification results in a 30–40% increase in the throughput of the UPDATE command as shown in Table 7 below.

Table 7. Throughput for various commands with the pipelined execution of UPDATE.

Command	Hops	Execution Times (ns)	Throughput (commands/sec)
INSERT	20	367	2724.79×10^3
INSERT	30	361	2770.08×10^3
INSERT	40	367	2724.79×10^3
UPDATE	20	189	5290.703×10^3
UPDATE	30	218	4587.4×10^3
UPDATE	40	224	4464.330×10^3
DELETE	20	153	6535.9×10^3
DELETE	30	162	6172.8×10^3

9. Conclusions and Future Work

In this study, we have implemented the order processing block of a trading system with FPGA technology. By migrating the functionality of order processing from software to hardware, we were able to obtain more than 2X of an advantage in throughput and order processing latency was reduced to less than 500 ns. The design was implemented with HLS. HLS methodology is comparatively new and is an emerging technology that is not mature as of yet. However, our observation is that the results of latency and throughput obtained with HLS are very close to Verilog implementation. With HLS, we achieved almost 4X–5X of an improvement in throughput for the INSERT and UPDATE commands compared to software implementation. However, to obtain results close to a highly optimized and efficient Verilog implementation, various optimization techniques need to be tried out as recommended below:

- Using HLS stream variables internally to implement FIFOs and carry out concurrent/overlapped executions of subfunctions of the three commands.
- Using an optimal mix of Verilog and C code in which certain latency and time-critical subfunctions are coded in Verilog, and the rest of the logic is coded in C and implemented in HLS.
- Design under test (DUT) consists of the Verilog implementation of the order processing block. As an alternative approach, the same DUT can be ported on an Intel HLS Compiler, and the results compared with those obtained from Xilinx Vivado HLS.

Author Contributions: Conceptualization, S.P. and S.R.; methodology, S.P.; software, M.B.; validation, S.P. and M.B., formal analysis, R.P.; writing—original draft preparation, S.P.; writing—review and editing, S.P. and M.B.; supervision, R.P. All authors have read and agreed to the published version of the manuscript.

Funding: Research received no external funding.

Institutional Review Board Statement: Not Applicable.

Informed Consent Statement: Not Applicable.

Data Availability Statement: Study does not report any data.

Conflicts of Interest: Authors declare no conflict of interest.

References

1. Top 10 Stock Exchanges in the World 2022. Available online: https://www.edudwar.com/top-10-stock-exchanges-in-the-world/ (accessed on 3 January 2022).
2. Readler, B. *Verilog by Example: A Concise Introduction for FPGA Design*; Full Arc Press: Washington, DC, USA, 2011.
3. Coussy, P.; Morawiec, A. *High-Level Synthesis: From Algorithm to Digital Circuit*; Springer: Berlin/Heidelberg, Germany, 2008.
4. Baranov, S. *High Level Synthesis of Digital Systems: For Data Path and Control Dominated Systems*; ISBN Canada: Toronto, ON, Canada, 2018.
5. Gajski, D.D.; Ramachandran, L. Introduction to high-level synthesis. *J. IEEE Des. Test Arch.* **1994**, *11*, 44–54. [CrossRef]
6. Ren, H. A brief introduction on contemporary High-Level Synthesis. In Proceedings of the IEEE International Conference on IC Design & Technology, Austin, TX, USA, 28–30 May 2014.
7. Sarkar, S.; Dabral, S.; Tiwari, P.K.; Mitra, R.S. Lessons and Experiences with High-Level Synthesis. *IEEE Des. Test Comput.* **2009**, *26*, 34–45. [CrossRef]
8. Vivado Overview. Available online: https://www.xilinx.com/products/design-tools/vivado/integration/esl-design.html (accessed on 3 January 2022).
9. Intel® High Level Synthesis Compiler. Available online: https://www.altera.com/products/design-software/high-level-design/intel-hls-compiler/overview.html (accessed on 3 January 2022).
10. C++/SystemC Synthesis. Available online: https://www.mentor.com/hls-lp/catapult-high-level-synthesis/c-systemc-hls (accessed on 4 February 2022).
11. Boutros, A.; Grady, B.; Abbas, M.; Chow, P. Build fast, trade fast: FPGA-based high-frequency trading using high-level synthesis. In Proceedings of the 2017 International Conference on ReConFigurable Computing and FPGAs (ReConFig), Cancun, Mexico, 4–6 December 2017.
12. Brogaard, J.A. High Frequency Trading and its Impact on Market Quality. In Proceedings of the 5th Annual Conference on Empirical Legal Studies, New Haven, CT, USA, 5–6 November 2010.
13. Chlistalla, M. *High-Frequency Trading Better than Its Reputation? Deutsche Bank Research Report*; Deutsche Bank: Frankfurt, Germany, 2011.
14. Chiu, J.; Lukman, D.; Modarresi, K.; Velayutham, A. *High Frequency Trading*; Stanford University Research Report; Stanford University: Stanford, CA, USA, 2011.
15. Leber, C.; Geib, B.; Litz, H. High Frequency Trading Acceleration Using FPGAs. In Proceedings of the 21st International Conference on Field Programmable Logic and Applications, Chania, Greece, 5–7 September 2011; Available online: https://ieeexplore.ieee.org/document/6044837 (accessed on 4 February 2022).
16. Malazgirt, G.A.; Sönmez, N.; Yurdakul, A. High Level Synthesis Based Hardware Accelerator Design for Processing SQL Queries. Available online: https://www.researchgate.net/publication/282503089_High_Level_Synthesis_Based_Hardware_Accelerator_Design_for_Processing_SQL_Queries (accessed on 4 February 2022).
17. Ruiz, M.; Sidler, D.; Sutter, G.; Alonso, G.; López-Buedo, S. Limago: An FPGA-Based Open-Source 100 GbE TCP/IPStack. *IEEE Trans. Electron Devices* **1988**, *35*, 2454–2455. Available online: https://ieeexplore.ieee.org/document/8891991 (accessed on 4 February 2022).
18. Kerrisk, M. *The Linux Programming Interface: A Linux and UNIX System Programming Handbook*, 1st ed.; No Starch Press: San Francisco, CA, USA, 2010.
19. Bovet, D.P.; Cesati, M. *Understanding the Linux Kernel*, 3rd ed.; O'Reilly Media: Sebastopol, CA, USA, 2005.
20. Market Infrastructure Business Development. Available online: https://www.lseg.com/areas-expertise/technology/capital-markets-technology-services/millennium-exchange (accessed on 4 February 2022).
21. Project Brainwave. Available online: https://www.microsoft.com/en-us/research/project/project-brainwave/ (accessed on 4 February 2022).
22. Swarm64. Available online: https://swarm64.com/ (accessed on 4 February 2022).
23. PostgreSQL: The World's Most Advanced Open Source Relational Database. Available online: https://www.postgresql.org/ (accessed on 4 February 2022).
24. Alveo U250 Data Center Accelerator Card. Available online: https://www.xilinx.com/products/boards-and-kits/alveo/u250.html (accessed on 4 February 2022).
25. BlackLynx. Available online: https://www.ryft.com/ (accessed on 4 February 2022).
26. Hsiue, K.D. FPGA-Based Hardware Acceleration for a Key-Value Store Database. Master's Thesis, Department of Electrical Engineering and Computer Science, Massachusetts Institute of Technology, Cambridge, MA, USA, 2014.
27. Papaphilippou, P.; Luk, W. Accelerating Database Systems Using FPGAs: A Survey. In Proceedings of the 28th International Conference on Field Programmable Logic and Applications (FPL), Dublin, Ireland, 27–31 August 2018.

Disclaimer/Publisher's Note: The statements, opinions and data contained in all publications are solely those of the individual author(s) and contributor(s) and not of MDPI and/or the editor(s). MDPI and/or the editor(s) disclaim responsibility for any injury to people or property resulting from any ideas, methods, instructions or products referred to in the content.

electronics

MDPI

Article

A Model of Thermally Activated Molecular Transport: Implementation in a Massive FPGA Cluster

Grzegorz Jabłoński [1,*], Piotr Amrozik [1] and Krzysztof Hałagan [2]

1 Department of Microelectronics and Computer Science, Lodz University of Technology, 93-005 Łódź, Poland
2 Department of Molecular Physics, Lodz University of Technology, 90-924 Łódź, Poland
* Correspondence: grzegorz.jablonski@p.lodz.pl

Abstract: In this paper, a massively parallel implementation of Boltzmann's thermally activated molecular transport model is presented. This models allows taking into account potential energy barriers in molecular simulations and thus modeling thermally activated diffusion processes in liquids. The model is implemented as an extension to the basic Dynamic Lattice Liquid (DLL) algorithm on ARUZ, a massively parallel FPGA-based simulator located at BioNanoPark Lodz. The advantage of this approach is that it does not use any exponentiation operations, minimizing resource usage and allowing one to perform simulations containing up to 4,608,000 nodes.

Keywords: distributed system; reconfigurable system; FPGA; ARUZ; Boltzmann statistics; molecular simulation

Citation: Jabłoński, G.; Amrozik, P.; Hałagan, K. A Model of Thermally Activated Molecular Transport: Implementation in a Massive FPGA Cluster. *Electronics* **2023**, *12*, 1198. https://doi.org/10.3390/electronics12051198

Academic Editors: Juan M. Corchado and Stefano Ricci

Received: 18 January 2023
Revised: 8 February 2023
Accepted: 28 February 2023
Published: 2 March 2023

Copyright: © 2023 by the authors. Licensee MDPI, Basel, Switzerland. This article is an open access article distributed under the terms and conditions of the Creative Commons Attribution (CC BY) license (https://creativecommons.org/licenses/by/4.0/).

1. Introduction

Computer simulations have become one of the most important research methods for non-equilibrium processes in chemistry and physics. The applied simulation techniques, however, have encountered difficulties with different types of problems, related to spatial and time scales, which are necessary to bring the system to a state of full equilibrium. Such phenomena are particularly interesting wherever the enthalpy factor (non-covalent interactions between molecules and atoms) is important (e.g., in soft matter, simple and complex liquids, and polymer solutions), as well as when temperature is one of the parameters of the studied phenomenon. The commonly used computational methods for non-equilibrium problems, such as phase separation, include nonlinear diffusion equation solvers [1,2], molecular dynamics methods [3], and stochastic Monte Carlo (MC) methods [4,5].

A particularly interesting method belonging to the stochastic category is the Dynamic Lattice Liquid (DLL) model [6–8], as it allows observing not only steady state static behavior but also the process of reaching equilibrium. It is based on the concept of cooperative motion of objects (elements). The positions of the elements are limited to the nodes of a face-centered cubic (FCC) lattice (coordination number $Z = 12$) for simplicity. The FCC lattice is commonly chosen in 3D as it has the highest coordination number among regular ones. The algorithm works on a completely occupied lattice, where the elements cannot easily move over a long distance due to the occupation of all neighboring lattice sites. In this case, the only way to move the elements with the excluded volume preserved is cooperative motion. In DLL, cooperative rearrangements take the form of closed loops of displacements that involve at least three elements (see Figure 1). Loops are formed spontaneously in a random way. The DLL model fulfills the continuity equation and provides the correlated movements of molecules as a model of real liquids. A discussion of the detailed balance and ergodicity in the DLL model was presented in [9].

This basic version of the DLL algorithm (also called the LOOPS mechanism here) models Brownian diffusion in a long time limit for simple liquids. However, to model more complex phenomena, the DLL model can be extended with additional functionalities, namely the so-called "mechanisms":

- The introduction of molecular bonds with excluded volume between elements must involve the BONDS mechanism, which is responsible for movement restriction related to the length-constant unbreakable bonds.
- The mechanisms BOND_BINDS and BOND_BREAKS are used when one wants to simulate the macromolecular polymerization and degradation processes, respectively.
- A growing macromolecule (in the case of polymerization, where new elements are joined to the molecule with some probability) can be terminated randomly using the TERMINATION mechanism.
- Chemical reactions of different orders can also be modeled with the REACTION mechanism, where elements can change their type with a given probability.
- Local trapping can be modeled with the MOBILITY mechanism, where movement of a given element can be restricted (e.g., due to its spatial position in the simulation box).
- Vector fields can be modeled using the VECTOR and REORIENTATION mechanisms.
- In the case where vacancies are present, the WAYS mechanism is used to model cooperative motion involving empty lattice nodes, forming a cooperative set of elements (chain-like and not necessarily a loop).
- The APERIODIC mechanism is used to build immobile obstacles such as walls.
- Thermal noise can be reduced in simulations by lowering the temperature with the ENERGY mechanism (introducing potential energy barriers).

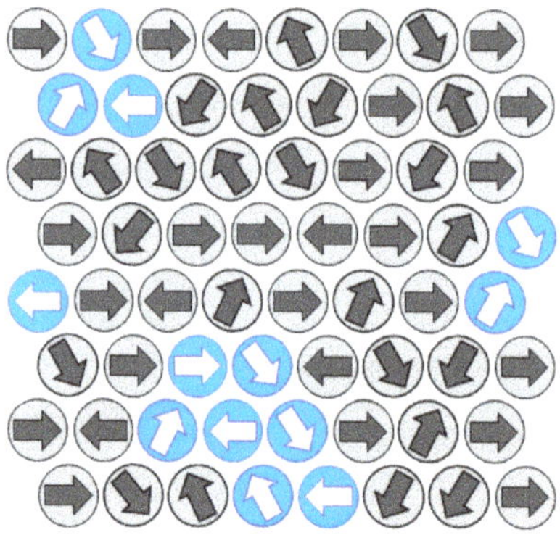

Figure 1. Attempts of movement in the DLL algorithm. Successful ones are marked as empty arrows in blue. A 2D case with $Z = 6$ is shown for clarity.

All the above mechanisms can be defined for a given type of element or spatial position in the simulation box, enabling modeling of various external fields.

The DLL model can be implemented efficiently with good scalability on a parallel machine equipped with low-latency communication interfaces to the nearest neighbors. A specialized field-programmable gate array (FPGA)-based simulator—ARUZ—was built for DLL simulations and has achieved performance orders of magnitude better than implementations on a sequential computer (see Table 4 in [10]).

In this paper, the details of Boltzmann's thermally activated molecular transport model [11] implemented as a DLL extension on ARUZ (ENERGY mechanism) are presented. The model allows taking into account potential energy barriers in molecular simulations and thus allows modeling of thermally activated diffusion processes in liquids. This approach does not use any exponentiation operations, allowing one to perform simulations containing 4,608,000 nodes, reaching 69 percent of the maximum simulation size achievable using only the LOOPS mechanism.

The performed simulations confirm that the implementation gives exact results compared with the values calculated theoretically.

Although the presented implementation works as an extension of the DLL algorithm, it is also applicable to other molecular movement models, especially those based on the lattice approach with high occupancy, such as direct exchange dynamics [12] or vacancy transport [13].

2. The ARUZ Simulator

The ARUZ [10], commissioned at the end of 2015 at Lodz Technopark (currently BioNanoPark), came as a result of close cooperation between the Department of Molecular Physics and the Department of Microelectronics and Computer Science, both from the Lodz University of Technology, complemented by the professional management expertise of the Ericpol (currently Ericsson) company. It is the first instance of a simulator built using TAUR technology [14]. This machine was designed to reflect the DLL algorithm in its hardware [15].

The ARUZ simulator consists of 2880 simulation boards called daughter boards (DBoards), interconnected by ca. 75,000 cables (see Figure 2). Each of them carries nine FPGAs: eight of them being called DSlaves (Artix XC7A200T), which constitute the resources for the nodes of the simulation algorithm, and the remaining one called DMaster (Zynq XC7Z015), which manages the operation of the DSlaves. The entire ARUZ thus contains 23,040 DSlaves. Each of the DSlaves can host up to a few hundred simulation nodes, depending on which features of the computational model are selected, and has communication interfaces to the eight closest neighboring FPGAs in a 3D simulation space.

Figure 2. Inside ARUZ. DBoards on interconnected panels.

Internally, each FPGA implements a grid of specialized processing cells dedicated to performing consecutive steps and complex calculations of the Dynamic Lattice Liquid algorithm. Assuming that every FPGA can host ca. 300 DLL cells [10], the entire simulation space consists of about 6.9 million nodes in the case of the basic DLL version.

3. Thermally Activated Diffusion Model

In the DLL algorithm, the individual molecules try to move in random directions, and the movement is possible only if the set of molecules is able to form a cooperative loop, where excluded volume interactions are naturally accounted for. The thermally activated diffusion model introduces an additional temperature-dependent restriction on molecule movement due to their interaction with neighbors. Such simulations have been performed previously only on a sequential computer [16–18], and no high-performance parallel implementation of this model is known. Different models can be considered in this case, including the kinetic MC test [16] based on the present state, forward-testing based on the next state, Metropolis sampling [4], or Glauber dynamics [19]. The first one was selected for the sake of simplicity and high performance. Additionally, the rest of the models take into account the forward system configuration, which would require a reversing phase of the simulation (going back to the starting point if the test fails), which complicates the parallel architecture a lot. Only the nearest neighbors were taken into account to simplify the interconnection topology and limit the number of transmission phases, but this simplification is not really significant in dense systems where long-distance interactions are mostly shielded by the rest of the system.

The ENERGY algorithm computes the probability of molecule movement in the following simulation phase (cooperative loops). Based on this probability, a pseudo-random number generator determines if the given molecule is immobilized in the current DLL cycle.

The involvement of nearest-neighbor interactions (or, in other words, temperature-dependent attraction or repulsion) must take into account the probability test related to the system energy. The system Hamiltonian for the ith lattice site populated by the X-type element is defined as follows:

$$\frac{E_i}{k_B T} = \frac{1}{k_B T}\left(\frac{1}{2}\sum_j \varepsilon_{XY_j} + H_{X_i}\right) \tag{1}$$

where the sum extends over all nearest neighbors and j can take any type Y. The following definitions apply:

- T is the absolute temperature, and k_B is the Boltzmann constant.
- H_X/k_BT is the interaction energy of the type X with the external field and can depend on the spatial position to model, for example, the temperature gradient.
- ε_{XY}/k_BT is the interaction energy of the i,j pair and is position-independent. In the presented model, ε_{XY} always equals ε_{YX}.

Both ε_{XY}/k_BT and H_X/k_BT are input data. For example, if four interactions ε_{XY} between elements are defined, with $X = A$ placed in the currently analyzed ith lattice node and $Y = A, B, C, D$, then

$$\begin{aligned}\frac{E_i}{k_B T} = \frac{1}{2k_B T}(\varepsilon_{AA}n_{AA} + \varepsilon_{AB}n_{AB} + \\ +\varepsilon_{AC}n_{AC} + \varepsilon_{AD}n_{AD}) + \frac{H_{A_i}}{k_B T}\end{aligned} \tag{2}$$

with multiplicities of i,j pairs $n_{AA} = 2$, $n_{AB} = 2$, $n_{AC} = 1$, and $n_{AD} = 1$, as illustrated in Figure 3.

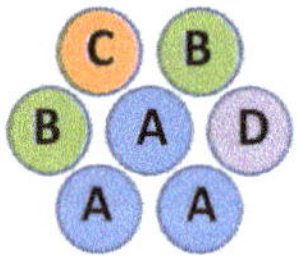

Figure 3. Example local configuration of interacting types shown in 2D.

A kinetic Monte Carlo test, representing the Boltzmann statistics, is applied (i.e., an effective attempt of motion is performed with a probability proportional to the energy of the current local state) [16,20]:

$$P_i = e^{-\frac{E_i}{k_B T}} \tag{3}$$

The test is executed for all elements in a given loop of possible cooperative movement, and all elements must pass it; otherwise, the whole loop is immobilized (when $exp(-E_i/k_BT) < random[0,1)$).

As a result, if $\varepsilon_{XY} > 0$, then an attractive interaction is present for the i,j pair. Interactions become effective at a finite temperature by reducing the probability of motion. Here,

ε_{XY} describes the barrier which has to be overcome in order to release contact between X and Y during the thermally activated diffusion process.

The average interaction energy per element is equal to $< E >= 1/N \sum_{i=1}^{N} E_i$, with N being the total number of elements in the lattice.

Note that the commonly known Metropolis sampling algorithm is not used because the loops have a spatial extent (over large distances, sometimes even 50 lattice constants [10]), and the test defined above involves the nearest neighbors only [21].

4. Implementation Requirements

To implement the kinetic MC test, the types of all the neighbors must be known. These are obtained using the local communication, described in detail in [10] (see Figure 11 therein for details about latency). After this information is acquired, the computations are performed independently in all nodes. The DLL algorithm's performance is mainly determined by the performance of the loop detection phase [22]. The time of a single cycle of the LOOPS algorithm amounts to ca. 100 microseconds. Therefore, in implementation of the ENERGY mechanism, the minimization of resource utilization and not the lowest latency is the main objective.

Implementing the test in a digital circuit requires choosing the number storage format and the required precision. Equation (3) contains the exponentiation, which is a quite complex operation that is difficult to implement in hardware.

By substituting (1) into (3), we obtain

$$P_i = e^{-(\frac{1}{2}\sum_{k=1}^{Z} \frac{\varepsilon_{XY}}{k_B T}(k) + H_X(x,y,z))} \tag{4}$$

where Z is the number of neighbors, which is assumed to be 12. If we define

$$\varepsilon_w(X,Y) = e^{(-\frac{1}{2}\frac{\varepsilon_{XY}}{k_B T})} \text{ and } \varepsilon_e(X) = e^{(-H_X(x,y,z))} \tag{5}$$

then we obtain

$$P_i = \prod_{k=1}^{Z} \varepsilon_w(X, Y_k) \cdot \varepsilon_e(X) \tag{6}$$

As $\varepsilon_w(X, Y_k)$ and $\varepsilon_e(X)$ are constant, no exponentiation operation is needed in the FPGA fabric, as they can be precomputed in the software.

For

$$y = e^x \tag{7}$$

we have

$$dy = e^x dx \tag{8}$$

Thus, we have

$$\frac{dy}{y} = \frac{e^x}{y} dx \tag{9}$$

and

$$\frac{dy}{y} = dx \tag{10}$$

Therefore, the absolute precision of x is equal to the relative precision of y. Therefore, to store ε_w and ε_e and perform operations on them, we need to apply a floating-point format, as it ensures a constant relative precision for the full range of stored values.

Because the assumed precision of the expression in the exponent in Equation (4) is 10^{-5}, and $log_2(10^{-5})$ is -16.7, to store ε_w and ε_e, we need a 17 bit mantissa.

Assuming that $\frac{1}{2}\frac{\varepsilon_{XY}}{k_B T}(k)$ in $H_X(x,y,z)$ has a range -10–10, from Equation (4), we can infer that P_i can vary in the range $e^{-10\cdot(Z+1)}$–$e^{10\cdot(Z+1)}$, which is thus $e^{-10\cdot 13}$–$e^{10\cdot 13}$, $2^{-187.55}$–$2^{187.55}$, and $2^{-2^{7.55}}$–$2^{2^{7.55}}$. This implies 9 bits for the floating point number exponent (8 for numbers greater than 1 and 8 for numbers smaller than 1) to encompass the entire range presented above.

In the vast majority of cases, the maximum number of simulation time steps is assumed to be 10^9. The mean probability of element movement is close to 6% for the FCC lattice [23] in the case of a basic version of DLL (additional mechanisms can only decrease it). Therefore, the number of steps in which the energy test will be used is approximately $0.06 \times 10^9 = 6 \times 10^7$ per element, so it will be necessary to perform the operation with probabilities in the order of $1/(6 \times 10^7)$. $Log_2(6 \times 10^7) = 25.8$, and therefore we need at least a 26 bit pseudo-random number generator. Please note that we do not need to compute a logarithm of the pseudo-random number as opposed to, for example, the solution presented in [24].

5. Implementation on FPGA

As the number of simulation nodes in DSlave (Artix XC7A200T) is limited to ca. 300, the optimization of the amount of hardware resources used by each node becomes a crucial factor. As can be seen in Figure 16 in [10], the most limiting resources are look-up tables (LUTs). Therefore, implementation using other available FPGA resources is desirable. The kinetic Monte Carlo test requires $Z + 1$ multiplications of ε_w and ε_e, as presented in Equation (6). This can be performed effectively by employing specialized hardware resources available in DSlave instead of configurable logic blocks (CLBs) [25], such as block random access memory (BRAM) [26] to store coefficients ε_w and ε_e and digital signal processing (DSP) slices [27] to perform multiplications.

Artix XC7A200T has 13,140 kb of BRAM [28] and 740 DSP slices, each containing a 25×18 multiplier. To implement multiplications in Equation (6), a single Xilinx Floating-Point Operator IP core [29] was used. This multiplier needs two DSP slices, limiting the number of nodes that can be implemented in one DSlave to 370. The block diagram of a module implementing the kinetic Monte Carlo test is presented in Figure 4.

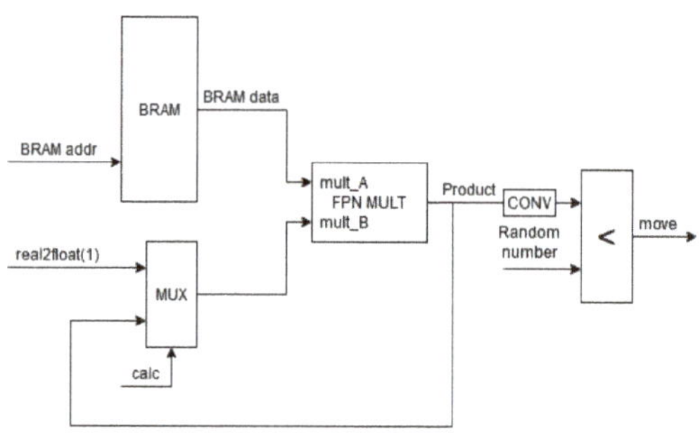

Figure 4. Block diagram of a module that implements the kinetic Monte Carlo test.

To reduce the number of CLBs, a special method for addressing BRAM is used which allows the concatenation of address vectors instead of employing more complicated calculations. The memory address is a concatenation of the following (see Table 1):

- An "e_offset" bit indicating parts of the memory storing ε_w and ε_e;
- A vector representing the type of a neighbor element "other_type";
- A vector representing the type of the considered element "my_type" (occupying the right-most bits).

The coefficient memory is divided into two sections. The first section of the memory stores ε_w coefficients and requires M^2 memory locations (assuming that ε_w occupies one location), where M is the number of types rounded up to the nearest power of two. The second section of the memory stores ε_e coefficients and requires M memory locations (assuming that ε_e occupies one location). Therefore, the total number of memory locations is $M^2 + M$.

Table 1. Example of BRAM utilization. White background denotes entries that are unused when only three element types are present.

Address (e_offset & other_type & my_type)	Data
0 & 00 & 00	$\varepsilon_w(A, A)$
0 & 00 & 01	$\varepsilon_w(A, B)$
0 & 00 & 10	$\varepsilon_w(A, C)$
0 & 00 & 11	$\varepsilon_w(A, D)$
0 & 01 & 00	$\varepsilon_w(B, A)$
0 & 01 & 01	$\varepsilon_w(B, B)$
0 & 01 & 10	$\varepsilon_w(B, C)$
0 & 01 & 11	$\varepsilon_w(B, D)$
0 & 10 & 00	$\varepsilon_w(C, A)$
0 & 10 & 01	$\varepsilon_w(C, B)$
0 & 10 & 10	$\varepsilon_w(C, C)$
0 & 10 & 11	$\varepsilon_w(C, D)$
0 & 11 & 00	$\varepsilon_w(D, A)$
0 & 11 & 01	$\varepsilon_w(D, B)$
0 & 11 & 10	$\varepsilon_w(D, C)$
0 & 11 & 11	$\varepsilon_w(D, D)$
1 & 00 & 00	$\varepsilon_e(A)$
1 & 00 & 01	$\varepsilon_e(B)$
1 & 00 & 10	$\varepsilon_e(C)$
1 & 00 & 11	$\varepsilon_e(D)$

In Table 1 an example of memory addressing is presented:

- Example 1: Four element types are used (types A, B, C, and D coded by a two-bit vector: A = 00, B = 01, C = 10, and D = 11). The address of an appropriate ε_w is a concatenation of the "e_offset" bit set to 0, and two two-bit vectors (representing the type of the considered element and a neighbor, respectively). The address of ε_e is the concatenation of a constant M^2 coded by a three-bit vector (taking bits of "e_offset" and "other_type" vectors) and a two-bit vector representing the type of element considered. In this example, all memory locations are used, and 16 of them are needed for ε_w and 4 for ε_e, leading to a total of 20.
- Example 2: Three element types are used (types A, B, and C coded by a two-bit vector: A = 00, B = 01, and C = 10). The addresses of the appropriate coefficients ε_w and ε_e are determined in the same way as in the previous example. Only gray-colored memory locations are used, but the required number of memory locations is still 20.

As the presented example shows, some parts of the memory are unused in some scenarios, and the memory utilization is higher than would be expected based on the number of types, but calculation of the memory address is kept very simple, and its implementation costs less CLBs for one node.

The FPN MULT block (see Figure 4) is 27×27 (1 bit for the sign, 10 bits for the exponent, and 16 bits for the mantissa) floating-point multiplier. The BRAM data are fed to the mult_A input of the multiplier. The data on the mult_B input are multiplexed. During the first cycle of calculations, the value of one represented in the floating-point format, and during the following cycles, the current product is fed. In this way, the result accumulates, giving Equation (6) after $Z + 1 = 13$ multiplication cycles.

Figure 5 presents the results of an example simulation that shows the module's operation. Its settings are as follows: a single type A surrounded by a homogeneous mixture of types B and C, $\varepsilon_{AB}/k_B T = 0.2$, and $\varepsilon_{AC}/k_B T = 0.01$. Thus, $e^{\left(-\frac{1}{2}\cdot\Sigma_{k=1}^{\frac{Z}{2}} 0.2 - \frac{1}{2}\cdot\Sigma_{k=1}^{\frac{Z}{2}} 0.01\right)} = \prod_{k=1}^{\frac{Z}{2}} e^{\left(-\frac{1}{2}\cdot 0.2\right)} \cdot \prod_{k=1}^{\frac{Z}{2}} e^{\left(-\frac{1}{2}\cdot 0.01\right)} = \prod_{k=1}^{\frac{Z}{2}} 0.90483\ldots \cdot \prod_{k=1}^{\frac{Z}{2}} 0.99501\ldots = 0.53255\ldots$

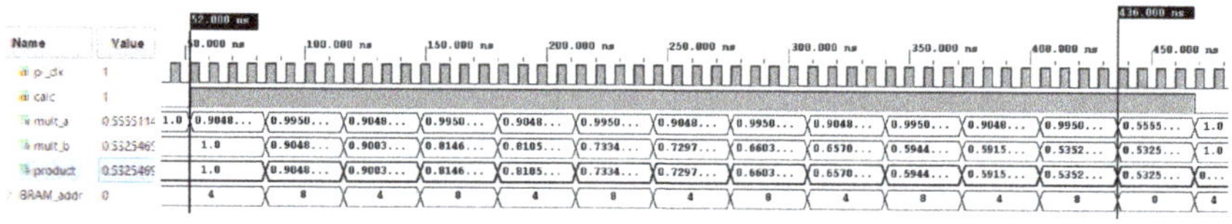

Figure 5. Result of the simulation.

In this simulation, one floating-point multiplication takes 4 clock cycles, so the whole operation takes $12 \cdot 4 = 48$ clock cycles. The clock period is 8 ns.

In the example, the neighbor nodes are placed evenly (homogeneous mixture). Thus, the BRAM address takes two values: four (0100), where $\varepsilon_w(B, A)$ is stored, and eight (1000), where $\varepsilon_w(C, A)$ is stored. The values read from BRAM are 0.90483... and 0.99501..., respectively. The result is 0.5325469..., which is correct.

The output of "FPN MULT" is compared with a random number. This random number is a 32 bit vector generated by a dedicated linear feedback shift register (LSFR). To make this comparison possible, the product is converted to a 32 bit fixed-point number. Both numbers being compared are treated as numbers in the range $[0, 1)$. The flag "move" is set to one when the "product" is greater than the generated random number.

There are two parameters of the floating-point multiplier core affecting the resource utilization and timing closure that can be adjusted: latency and DSP usage. Latency can be configured to between 0 and 8 clock cycles. The multiplier can use zero ("no DSP"), one ("full DSP" setting), or two ("max DSP") DSP48 blocks per instance.

Table 2 summarizes the results of exploration of a design space and shows the overhead of adding the ENERGY mechanism to the simulation. The results for the maximum number of nodes that can be placed in a single FPGA that can be successfully implemented without timing issues are presented in the table.

Table 2. Exploration of a design space.

Nodes per Chip	Mechanisms	Multiplier Parameters	LUTs (%)	FFs (%)	BRAMs (%)	DSPs (%)
200	LOOPS	N/A	58.01	29.21	0.00	0.00
288	LOOPS	N/A	81.52	40.98	0.00	0.00
128	LOOPS, ENERGY	latency 3, no DSP	80.09	34.93	0.00	35.07
200	LOOPS, ENERGY	latency 3, full DSP	79.17	47.90	27.03	54.79
200	LOOPS, ENERGY	latency 3, max DSP	76.56	48.05	54.05	54.79
200	LOOPS, ENERGY	latency 4, max DSP	77.58	48.95	54.05	54.79
200	LOOPS, ENERGY	latency 8, max DSP	78.24	50.87	54.05	54.79

The maximum number of simulation nodes that can be placed in a DSlave is limited by the number of LUTs. It is typically not possible to route a design with more than 80% LUT utilization. Increasing the latency of the multiplier improves the timing closure but also slightly increases resource utilization. The timing closure is not possible with the multiplier latency set to one. Setting it to two results in successful implementation only for three out of all eight DSlaves on the board. It is possible to obtain consistently positive results for the synthesis with the latency set to at least three. At the "max DSP" setting, over half of the DSPs are used. As the only other mechanism that is able to use DSPs is the WAYS mechanism [22], and it uses only one DSP48 block per node, this is not a limiting factor. However, halving DSP usage by applying the "full DSP" setting does not cause a significant increase in LUTs. On the other hand, using the "no DSP" setting increases the LUT usage, enormously reducing the maximum number of nodes per chip to 128.

The inclusion of the ENERGY mechanism consumes ca. 20% of the LUTs in fully utilized FPGA. As a result, it decreases the maximum number of nodes per chip to 200 (4,608,000 in the entire machine) compared with 288 for the simulation using only LOOPS.

6. Example Simulation

The Hamiltonian in Equation (1) can model a kind of system called the conserved-order parameter [30] Ising model [31], or model B in the Hohenberg–Halperin classification [32]. The model assumes that the diffusion can be suppressed in X(Y)-type-rich regions ($\varepsilon_{XX(YY)}/k_BT > 0$) or at the X-Y interfaces [33] ($\varepsilon_{XY}/k_BT > 0$) as a result of nearest-neighbor interactions.

As the probability of any state in equilibrium is given by the Boltzmann distribution, the interaction ε_{XY} causes configurations where particles are clustered together to be low in energy and therefore more likely at low temperatures. The critical interaction value for the FCC lattice (defined with a critical temperature T_C) after which the systems undergo the order-disorder transition (second-order phase transition) is $\varepsilon/2k_BT_C = 0.204\ldots$ [34,35].

This kind of system was used to study many physical problems for which the kinetics of mixing or demixing matters (i.e., where diffusivity of elements is highly related to their neighborhood), such as for binary alloys [36,37], liquid mixtures [38,39], and polymer blends [40,41]. Obviously, if more than two types with many more pairs of interactions are defined, then the modeled system possesses much higher complexity than the simple Ising model, and its properties strictly depend on the defined conditions.

Figure 6 presents simulation snapshots for a box initially filled with two interacting types: X = A and Y = B (50%:50%). The system consisted of 3,538,944 nodes with periodic boundary conditions. In the first simulation step, the types were fully separated (see Figure 6a in all cases). Figure 6b presents the box configuration after $t = 10^5$ cycles of the algorithm and $\varepsilon_{AB}/2k_BT = 0$ (no interaction). The box was mixed by diffusive motion of the elements (by the LOOPS mechanism [10,22]), ending with the totally random configuration. When the interaction was set to $\varepsilon_{AB}/2k_BT = 0.1$ (Figure 6c), the mixing process was slowed because diffusive motion was limited in the areas where A and B were in contact. However, after $t = 10^6$, in this case, the system was again random. The application of $\varepsilon_{AB}/2k_BT = 0.5$ (Figure 6d) resulted in a stable separation in time because the critical value of the interaction for the FCC lattice was exceeded, and the system remained ordered. The introduction of $\varepsilon_{AB}/2k_BT$ does not restrict movement within regions rich in A or B ($\varepsilon_{AA}/2k_BT$ and $\varepsilon_{BB}/2k_BT$ were set to zero). In the investigated cases, ARUZ achieved a performance of 5260 cycles/s (18.6×10^9 lattice updates per second (LUPS) [10]) for $\varepsilon_{AB}/2k_BT = 0.01$ and up to 7300 cycles/s (25.8×10^9 LUPS) for $\varepsilon_{AB}/2k_BT = 0.1$ with a completely random initial configuration because the energy tests excluded $\tilde{4}6$% of all elements from the diffusive motion analysis in this case.

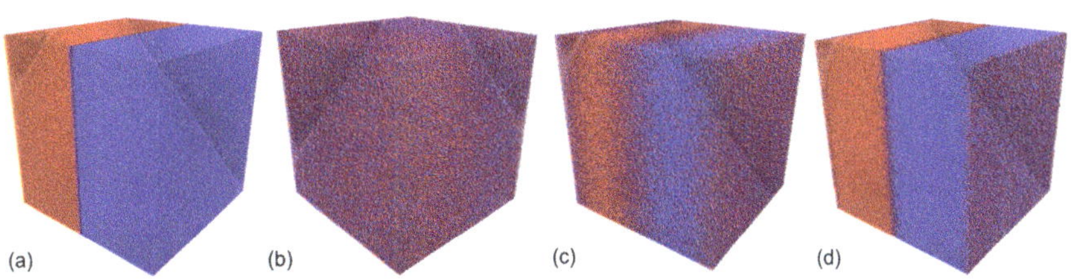

Figure 6. Example of simulation snapshots for (**a**) $t = 0$, (**b**) $\varepsilon_{AB}/2k_BT = 0$ and $t = 10^5$, (**c**) $\varepsilon_{AB}/2k_BT = 0.1$ and $t = 10^5$, and (**d**) $\varepsilon_{AB}/2k_BT = 0.5$ and $t = 10^6$. Types A and B are marked with different colors.

7. Conclusions

An efficient approach to FPGA-based simulation of Boltzmann's thermally activated diffusion was developed. It avoids the expensive exponentiation operation in the FPGA fabric by using precomputed values of the Boltzmann weights. A special method for

BRAM addressing was used, eliminating complicated calculations thanks to address vector concatenation. Application of the described model increased the performance of the simulation measured in LUPS, as the energy tests excluded a large portion of all elements from diffusive motion. The simulations performed confirmed that the implementation gives exact results compared with the theoretically calculated values.

The presented implementation is applicable to other lattice algorithms where Boltzmann weights need to be used.

Author Contributions: Conceptualization, K.H.; methodology, G.J., P.A. and K.H.; software, G.J. and P.A.; validation, P.A. and K.H.; formal analysis, G.J.; investigation, G.J., P.A. and K.H.; data curation, K.H.; writing—original draft preparation, G.J.; writing—review and editing, G.J., P.A. and K.H.; visualization, K.H.; supervision, G.J. All authors have read and agreed to the published version of the manuscript.

Funding: This research was funded by Polish National Science Centre grant UMO-2017/25/B/ST5/01110.

Data Availability Statement: Not avaliable.

Conflicts of Interest: The authors declare no conflict of interest.

Abbreviations

The following abbreviations are used in this manuscript:

ARUZ	Analyzer of Real Complex Systems (in Polish: Analizator Rzeczywistych Układów Złożonych)
BRAM	Block random access memory
DLL	Dynamic Lattice Liquid
FCC	Face-centered cubic
FPGA	Field-programmable gate array
LUPS	Latice updates per second
TAUR	Technology of Real Word Analyzers (in Polish: Technologia Analizatorów Układów Rzeczywistych)

References

1. Cahn, J.W.; Hilliard, J.E. Free energy of a nonuniform system. I. Interfacial free energy. *J. Chem. Phys.* **1958**, *28*, 258–267. [CrossRef]
2. Cahn, J.W. On Spinodal Decomposition. *Acta Metall.* **1961**, *9*, 795–801. [CrossRef]
3. Binder, K.; Ciccotti, G. *Monte Carlo and Molecular Dynamics of Condensed Matter*; Società Italiana di Fisica: Bologna, Italy, 1996.
4. Metropolis, N.; Rosenbluth, A.W.; Rosenbluth, M.N.; Teller, A.H.; Teller, E. Equations of State Calculations by Fast Computing Machines. *J. Chem. Phys.* **1953**, *21*, 1087–1092. [CrossRef]
5. Binder, K.; Heerman, D.W. *Monte Carlo Simulation in Statistical Physics. An Introduction*, 4th ed.; Springer: Berlin, Germany, 2002.
6. Pakuła, T.; Teichmann, J. *Model for Relaxation in Supercooled Liquids and Polymer Melts*; MRS Online Proceedings Library: Berlin/Heidelberg, Germany, 1996; Volume 455, p. 211. [CrossRef]
7. Polanowski, P.; Jeszka, J.K.; Matyjaszewski, K. Polymer brushes in pores by ATRP: Monte Carlo simulations. *Polymer* **2020**, *211*, 123124. [CrossRef]
8. Kozanecki, M.; Halagan, K.; Saramak, J.; Matyjaszewski, K. Diffusive properties of solvent molecules in the neighborhood of a polymer chain as seen by Monte-Carlo simulations. *Soft Matter* **2016**, *12*, 5519–5528. [CrossRef]
9. Pakula, T. Simulation on the completely occupied lattices. In *Simulation Methods for Polymers*; Marcel Dekker: New York, NY, USA; Basel, Switzerland, 2004.
10. Kiełbik, R.; Hałagan, K.; Zatorski, W.; Jung, J.; Ulański, J.; Napieralski, A.; Rudnicki, K.; Amrozik, P.; Jabłoński, G.; Stożek, D.; et al. ARUZ—Large-scale, massively parallel FPGA-based analyzer of real complex systems. *Comput. Phys. Commun.* **2018**, *232*, 22–34. [CrossRef]
11. Jabłoński, G.; Amrozik, P.; Hałagan, K. Molecular Simulations Using Boltzmann's Thermally Activated Diffusion—Implementation on ARUZ—Massively-parallel FPGA-based Machine. In Proceedings of the 2021 28th International Conference on Mixed Design of Integrated Circuits and System, Lodz, Poland, 24–26 June 2021; pp. 128–131. [CrossRef]
12. Kawasaki, K.; Ohta, T. Theory of Early Stage Spinodal Decomposition in Fluids near the Critical Point. II. *Prog. Theor. Phys.* **1978**, *59*, 362–374. [CrossRef]
13. Yaldram, K.; Binder, K. Spinodal decomposition of a two-dimensional model alloy with mobile vacancies. *Acta Metall. Mater.* **1991**, *39*, 707–717. [CrossRef]

14. Jung, J.; Kiełbik, R.; Hałagan, K.; Polanowski, P.; Sikorski, A. Technology of Real-World Analyzers (TAUR) and its practical application. *Comput. Methods Sci. Technol.* **2020**, *26*, 69–75.
15. Polanowski, P.; Jung, J.; Kielbik, R. Special Purpose Parallel Computer for Modelling Supramolecular Systems based on the Dynamic Lattice Liquid Model. *Comput. Methods Sci. Technol.* **2010**, *16*, 147–153. [CrossRef]
16. Pakula, T.; Cervinka, L. Modeling of medium-range order in glasses. *J. Non-Cryst. Solids* **1998**, *232–234*, 619–626. [CrossRef]
17. Halagan, K.; Polanowski, P. Kinetics of spinodal decomposition in the Ising model with Dynamic Lattice Liquid (DLL) dynamics. *J. Non-Cryst. Solids* **2009**, *355*, 1318–1324. [CrossRef]
18. Halagan, K.; Polanowski, P. Order-disorder transition in 2D conserved spin system with cooperative dynamics. *J. Non-Cryst. Solids* **2015**, *127*, 585–587. [CrossRef]
19. Glauber, R.J. Time-Dependent Statistics of the Ising Model. *J. Math. Phys.* **1963**, *4*, 294–307. [CrossRef]
20. Pakula, T. Collective dynamics in simple supercooled and polymer liquids. *J. Mol. Liq.* **2000**, *86*, 109–121. [CrossRef]
21. Hałagan, K. Investigation of Phase Separation and Spinodal Decomposition Phenomena with Cooperative Dynamics. Ph.D. Thesis, Lodz University of Technology, Łódź, Poland, 2013.
22. Kiełbik, R.; Hałagan, K.; Rudnicki, K.; Jabłoński, G.; Polanowski, P.; Jung, J. Simulation of diffusion in dense molecular systems on ARUZ—Massively-parallel FPGA-based machine. *Comput. Phys. Commun.* **2023**, *283*, 108591. [CrossRef]
23. Polanowski, P.; Pakula, T. Studies of mobility, interdiffusion, and self-diffusion in two-component mixtures using the dynamic lattice liquid model. *J. Chem. Phys.* **2003**, *118*, 11139–11146. [CrossRef]
24. Migacz, S.; Dutka, K.; Gumienny, P.; Marchwiany, M.; Gront, D.; Rudnicki, W.R. Parallel Implementation of a Sequential Markov Chain in Monte Carlo Simulations of Physical Systems with Pairwise Interactions. *J. Chem. Theory Comput.* **2019**, *15*, 2797–2806.
25. 7 Series FPGAs Configurable Logic Block. Available online: https://docs.xilinx.com/v/u/en-US/ug474_7Series_CLB (accessed on 17 January 2023).
26. 7 Series FPGA Memory Resources User Guide. Available online: https://docs.xilinx.com/v/u/en-US/ug473_7Series_Memory_Resources (accessed on 17 January 2023).
27. 7 Series DSP48E1 Slice User Guide. Available online: https://docs.xilinx.com/v/u/en-US/ug479_7Series_DSP48E1 (accessed on 17 January 2023).
28. 7 Series Product Selection Guide. Available online: https://www.xilinx.com/content/dam/xilinx/support/documents/selection-guides/7-series-product-selection-guide.pdf (accessed on 17 January 2023).
29. Floating–Point Operator v7.1 LogiCore IP Product Guide. Availiable online: https://www.xilinx.com/content/dam/xilinx/support/documents/ip_documentation/floating_point/v7_1/pg060-floating-point.pdf (accessed on 17 January 2023).
30. Newman, M.E.J.; Barkema, G.T. *Monte Carlo Methods in Statistical Physics*; Clarendon Press: Oxford, UK, 1999.
31. Ising, E. Beitrag zur Theorie des Ferromagnetismus. *Z. Physik* **1924**, *31*, 253. [CrossRef]
32. Hohenberg, P.C.; Halperin, B.I. Theory of dynamic critical phenomena. *Rev. Mod. Phys.* **1877**, *49*, 435. [CrossRef]
33. Marko, J.F.; Barkema, G.T. Phase ordering in the Ising model with conserved spin. *Phys. Rev. E* **1995**, *52*, 2522. [CrossRef] [PubMed]
34. Liu, A.J.; Fisher, M.E. The three-dimensional Ising model revisited numerically. *Physica A* **1989**, *156*, 35–76. [CrossRef]
35. Yu, U. Critical temperature of the Ising ferromagnet on the FCC, HCP, and DHCP lattices. *Physica A* **2015**, *419*, 75–79. [CrossRef]
36. Gaulin, B.; Spooner, S.; Morii, Y. Kinetics of phase separation in $Mn_{0.67}Cu_{0.33}$. *Phys. Rev. Lett.* **1987**, *59*, 668. [CrossRef]
37. Wagner, R. *Chapter 5 in Phase Transformations in Materials*; Wiley-VCH: Weinheimd, Germany, 2001.
38. Wong, N.; Knobler, C. Light-Scattering Studies of Phase Separation in Isobutyric Acid + Water Mixtures. 2. Test of Scaling. *J. Phys. Chem.* **1981**, *85*, 1972–1976.
39. Mauri, R.; Shinnar, R.; Triantafyllou, G. Spinodal decomposition in binary mixtures. *Phys. Rev. E* **1996**, *53*, 2613. [CrossRef]
40. Bates, F.; Wiltzius, P. Spinodal decomposition of a symmetric critical mixture of deuterated and protonated polymer. *J. Chem. Phys.* **1989**, *91*, 3258–3274. [CrossRef]
41. Demyanchuk, I.; Wieczorek, A.; Hołyst, R. Percolation-to-droplets transition during spinodal decomposition in polymer blends, morphology analysis. *J. Chem. Phys.* **2004**, *121*, 1141–1147. [CrossRef]

Disclaimer/Publisher's Note: The statements, opinions and data contained in all publications are solely those of the individual author(s) and contributor(s) and not of MDPI and/or the editor(s). MDPI and/or the editor(s) disclaim responsibility for any injury to people or property resulting from any ideas, methods, instructions or products referred to in the content.

electronics

MDPI

Article

Electromyogram (EMG) Signal Classification Based on Light-Weight Neural Network with FPGAs for Wearable Application

Hyun-Sik Choi

Department of Electronic Engineering, College of IT Convergence Engineering, Chosun University, Gwangju 61452, Republic of Korea; hs22.choi@chosun.ac.kr

Abstract: Recently, the application of bio-signals in the fields of health management, human–computer interaction (HCI), and user authentication has increased. This is because of the development of artificial intelligence technology, which can analyze bio-signals in numerous fields. In the case of the analysis of bio-signals, the results tend to vary depending on the analyst, owing to a large amount of noise. However, when a neural network is used, feature extraction is possible, enabling a more accurate analysis. However, if the bio-signal time series is analyzed as is, the total neural network increases in size. In this study, to accomplish a light-weight neural network, a maximal overlap discrete wavelet transform (MODWT) and a smoothing technique are used for better feature extraction. Moreover, the learning efficiency is increased using an augmentation technique. In designing the neural network, a one-dimensional convolution layer is used to ensure that the neural network is simple and light-weight. Consequently, the light-weight attribute can be achieved, and neural networks can be implemented in edge devices such as the field programmable gate array (FPGA), yielding low power consumption, high security, fast response times, and high user convenience for wearable applications. The electromyogram (EMG) signal represents a typical bio-signal in this study.

Keywords: bio-signals; artificial intelligence technology; light-weight neural network; maximal overlap discrete wavelet transform (MODWT); smoothing technique; augmentation technique; edge devices; field programmable gate array (FPGA); electromyogram (EMG)

Citation: Choi, H.-S. Electromyogram (EMG) Signal Classification Based on Light-Weight Neural Network with FPGAs for Wearable Application. *Electronics* **2023**, *12*, 1398. https://doi.org/10.3390/electronics12061398

Academic Editors: Andres Upegui, Andrea Guerrieri and Laurent Gantel

Received: 3 February 2023
Revised: 28 February 2023
Accepted: 1 March 2023
Published: 15 March 2023

Copyright: © 2023 by the author. Licensee MDPI, Basel, Switzerland. This article is an open access article distributed under the terms and conditions of the Creative Commons Attribution (CC BY) license (https://creativecommons.org/licenses/by/4.0/).

1. Introduction

Various bio-signals have been developed, such as electrocardiogram (ECG), electromyogram (EMG), photo-plethysmography (PPG), and electroencephalogram (EEG). Many artificial intelligence networks using these bio-signals are being developed, and the fields of application are diverse [1,2]. Typical application fields include health care, human–computer interaction (HCI), and user authentication [3–6]. In this study, the EMG signal was treated as a representative bio-signal. In the case of the EMG signal, the measurement is simple; therefore, it is expected to be used in fields such as user authentication and HCI [7,8]. EMG is measured by amplifying a small signal coming from the movement of the muscle, and the measurement equipment has three input terminals [9]. Two terminals were used to measure the signal in the middle and end of the muscle. The last terminal was used as a reference signal, which is far from the muscle. This is amplified by the differential amplification unit to determine muscle movement [10]. The EMG sensor is easy to attach and can be used for HCI by attaching it to the most active hand. For HCI applications that use EMG, the classification of the signals using artificial intelligence has been the subject of many studies [11,12]. The increasing use of bio-signals is due to the rapid progress of artificial intelligence and big data technology.

In the case of EMG signals, as with other bio-signals, various noise components are included in the measurement [13]. This creates a problem in that the analysis of results can differ depending on the analyst. However, when a neural network is used, feature extraction is possible and more accurate, and consistent results can be obtained [14]. In [15], based on EMG signals measured in 16 channels, support vector machine (SVM) and generalized regression neural network (GRNN) artificial intelligence algorithms were employed using the root mean square (RMS), an autoregressive model (AR), and the slope sign change (SSC). In this manner, the six hand gestures were classified with 98% accuracy. In [16], a user recognition study was conducted using the EMG signal measured when drawing an unlock pattern on a smartphone screen. The EMG signal was measured in the flexor digitorum superficialis (FDS) muscle of the forearm using OpenBCI. Using one-class SVM (OCSVM) and extracting features, such as the mean absolute value (MAV), variance (VAR) in the time domain, the user was recognized 98.2% of the time. In [17], a fast Fourier transform (FFT) was performed using EMG signals, and based on this, it was confirmed that diseases such as neuropathy muscle disease could be predicted. However, if the time series is used as is or if FFT is used, the neural network for feature extraction becomes significant, with more than 100,000 weight values, making it unsuitable for wearable applications [18,19]. Most of the heavy EMG signal analyses are implemented in a structure that transmits data to a server and analyzes them on the server. This means that many hardware resources are used in the server, and the power consumption is high. Moreover, security is emerging as a big problem in EMG analysis, while, regarding authentication, problems such as slow response times may occur [20,21].

To solve this problem, the field of bio-signal processing in edge devices has received considerable attention [22–24]. This is because when using an edge device, low power, a fast response speed, and security can be expected [25]. However, difficulty in achieving low power in the case of CPU- and GPU-based artificial intelligence systems is a problem. This is because their operation is clock-based and consumes considerable power in the process of accessing memory. To prevent this, a field programmable gate array (FPGA) with a distributed structure is a possible alternative, and low power and fast response times can be secured through various structure optimizations [26,27]. In the case of an edge device with a CPU/GPU or an embedded system, only one execution is performed, whereas an FPGA has the advantage of executing multiple instructions simultaneously. Therefore, FPGAs have received significant attention in the field of edge devices [28,29].

In the case of an inference accelerator using an FPGA, many studies have been conducted [30]. However, when an FPGA is used, external memory such as dynamic random access memory (DRAM) is required to store the weight values, and considerable energy is therefore consumed for reading memory processes. To prevent this, edge devices that are more compressed are being studied, which is also the focus of this study. This study focuses on hardware configurations that can be applied in real life. The hardware compression methods used are largely based on (1) compression of the model network itself using algorithms, (2) compression of the computation methods (e.g., MobileNet [31], a systolic array structure for the tensor processing unit (TPU) [32]), and (3) a pruning method that uses weight sparsity and bit quantization [33]. In this study, hardware compression was carried out by focusing on methods (1) and (3). Therefore, the main purpose is to design a light-weight neural network suitable for an edge device, such as an FPGA, capable of parallel operation and with low memory access. With such a neural network, memory components for storing weight values can exist inside the FPGA without any external DRAM memory access (e.g., ResNet minimum storage requirement for 26 million weight values of 104 MB).

For this purpose, a frequency-filtered signal is used by signal processing. The maximal overlap discrete wavelet transform (MODWT) and smoothing algorithm for this signal are used. Moreover, an augmentation technique is used to increase learning efficiency. For the neural network design, a one-dimensional convolution layer is used for light-weight systems that are suitable for wearable applications. The neural network implemented in this manner is deployed to the FPGA, which is an edge device, and performs an appropriate inference operation. Through this, a low-power, high-speed, and high-security system with an edge device can be implemented even in low-cost FPGAs for artificial intelligence. In particular, the size of the model network can be reduced using an efficient feature extraction method that is suitable for EMG signals; through this, operation with high accuracy and low power consumption without access to external memory is possible. Additionally, bit quantization was performed to secure the compression of the edge device. The main application area is wearable systems for HCI or user authentication. Section 2 investigates the signal processing of EMG signals, and Section 3 reports on the structure of the neural network. Section 4 shows the overall hardware structure and verification process through high level synthesis (HLS) for hardware deployment. Section 5 presents the conclusions, discussions, and future research directions.

2. Signal Processing

2.1. Data Augmentaion

For the EMG signal input, "sEMG for Basic Hand movements Data Set" of the existing UCI machine learning repository was used [34]. This is the measurement data for five people, and the hand gestures are composed of six; thus, the final goal is to classify the six movements using the measured EMG signals. The measurement sampling frequency was 500 Hz, and the measurement was performed using two channels. The sampling frequency was 500 Hz because the most important data in the case of EMG signals are included in the range from 10 to 250 Hz. However, because a wearable device is assumed in this study, classification was performed using only data from channel 1. This causes a decrease in accuracy, but because one channel is mostly used in the actual HCI environment, one channel is used. In a real environment, the six hand gestures would be difficult to distinguish with one channel of data. In actual wearable device applications, classification for approximately two to three classes is considered. Both the EMG sensor and artificial intelligence will be produced in the form of a smartwatch. The actual application fields through this will be HCI implementation for two or three hand motions, analysis of carpal tunnel syndrome, and user authentication. However, for comparison with other neural networks, the accuracy was obtained by performing the classification of all six hand gestures. Additionally, in the case of measurement data, the amount of data is insufficient because it is data from five people; however, the data were used only for the feasibility test of the proposed neural network. The dataset "EMG data for gestures Data Set" of 36 subjects and eight channels for six hand gestures were also verified with similar sequences. A 98% classification accuracy was observed for the validation data (not shown here). The increased accuracy is related to the increased channel numbers [35]. In the future, the EMG signal will be measured via its own compact EMG modules and will be used as a wearable device. A high-efficiency wearable device for EMG signal acquisition is currently under development. Figure 1a shows a prototype of the EMG sensor that is currently under development. A differential amplifier was used, and signals from 10 to 250 Hz were obtained using frequency filtering. Figure 1b shows the measured EMG signals of the wrist. In the future, the signals will be produced in the form of a wearable watch through miniaturization and low power consumption and will be combined with an edge device for artificial intelligence to detect wrist movements.

In the case of the EMG input dataset, because the number of datasets is small, it is unsuitable for training. Therefore, data augmentation was implemented using additive noise and magnitude warping. In the case of additive noise, Gaussian noise was added, and the entire signal was processed after normalization. The equations below describe the creation of a new signal by adding additive Gaussian noise [36].

$$x_i^* = x_i + n \tag{1}$$

$$n \sim Gaussian(\mu = 0, \sigma = \sqrt{\frac{x_i^2}{SNR}}) \tag{2}$$

where x_i is the original signal and x_i^* is the augmented signal. n is the additive noise component and Gaussian noise is randomly added according to the signal-to-noise ratio

(SNR). Figure 2 shows a representative EMG signal and an augmentation signal using the additive noise method.

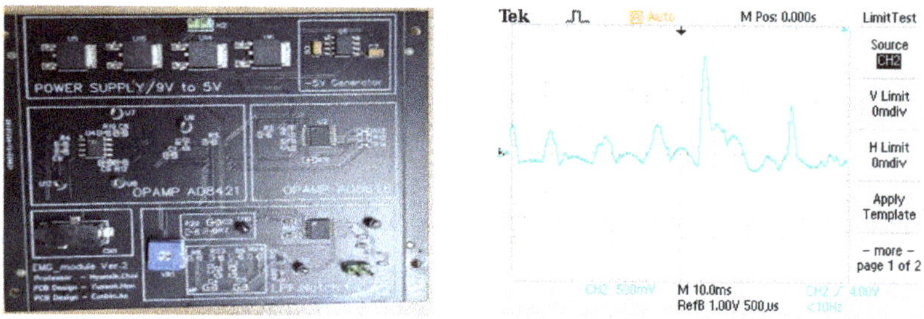

(**a**) (**b**)

Figure 1. (**a**) Prototype for electromyogram (EMG) sensor and (**b**) measurement results in the wrist.

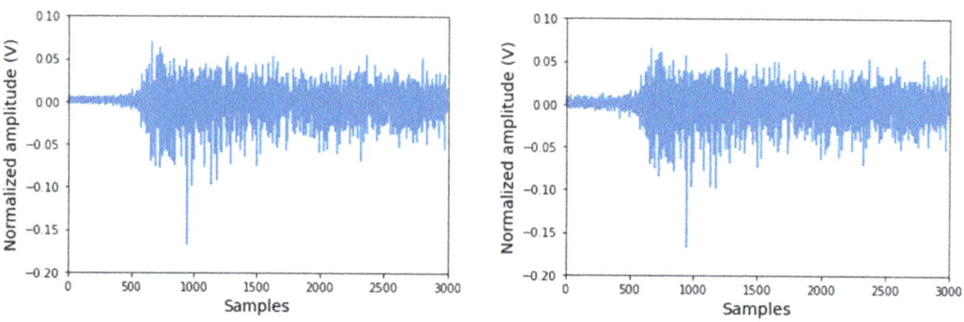

(**a**) (**b**)

Figure 2. Representative (**a**) EMG signal, (**b**) augmentation signal using additive noise of "sEMG for Basic Hand movements Data Set".

Moreover, the moving average, permutation, and magnitude warping techniques can be used for data augmentation. In this study, only the magnitude warping technique, which is known to be the most efficient, was used [36]. As a result of actual verification, the improvement in accuracy due to the moving average technique was insignificant, and in the case of the permutation technique, the accuracy is rather reduced. The magnitude warping method is shown in the following equations:

$$x_i^* = x_i \cdot CubicSpline(r) \tag{3}$$

$$\mathrm{r} = \{\mathrm{r}(t_1),\ \mathrm{r}(t_2),\ \ldots,\ \mathrm{r}(t_T)\} \tag{4}$$

where t is the sampling time. A random curve is generated using the CubicSpline method, which is an interpolated curve, and warping is performed on the magnitude. Through this augmentation method, training was performed by increasing the size of the dataset by approximately three times. This resulted in an increase in the classification accuracy of approximately 3% compared to when the data augmentation method was not used.

2.2. Filtering and Smoothing

The EMG signals were completely removed at 50 Hz and 0 Hz using an additional filter. They are signals that have passed through the band pass filter; however, when the signal analysis is performed through FFT, DC and 50 Hz components still exist; therefore, 50 Hz and 0 Hz signals were additionally removed using an additional high-performance filter for signal processing. The first filter was an active second-order high-pass filter with a cutoff frequency (f_c) of 10 Hz. The second filter was an active notch filter with f_c of 50 Hz. These are designed for easy hardware implementation in an EMG sensor system. The active notch filter is designed using the active twin-T notch filter method [37]. Figure 3 shows a representative signal before and after applying additional filters. This is displayed as a frequency spectrum for ease of comparison.

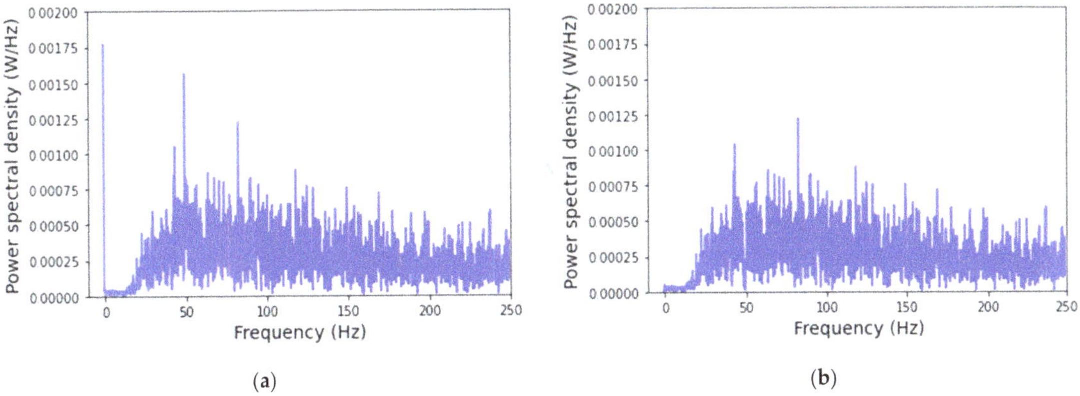

Figure 3. Frequency spectrum of EMG signal (**a**) before and (**b**) after the use of additional filters.

In addition to using the filtering method for the time series data, the effective noise components can be reduced through a smoothing technique. In this study, the Savitzky–Golay filter was used as the smoothing technique. Although several smoothing techniques exist, the Savitzky–Golay method was used because it is easy to perform for a simple hardware implementation. The Savitzky–Golay filter can mathematically replace smoothing using a polynomial regression model by providing a specific impulse response without calculating the regression model within the window of every time step in performing smoothing using a regression model. The time window was 42 ms, and a cubic equation was used for polynomial regression. Through the filtering method and smoothing technique, the accuracy could be improved by approximately 2%.

2.3. Feature Extraction

For feature extraction of the EMG signal, the MODWT method was used for easy implementation in the FPGA. The wavelet transform was developed to perform time and frequency domain analyses simultaneously. The wavelet transform has the advantage of being able to deal with information in the time domain instead of sacrificing some accuracy in the frequency domain. Among them, the discrete wavelet transform (DWT) based on orthonormal wavelet is frequently used; however, MODWT is more sensitive to circular shifts than the general DWT. This can be interpreted as a linear filter result. In hardware implementation, a finite impulse response (FIR) filter is used using digital logic. The above results indicate that MODWT, which is easy to interpret in combination with events that actually occur in nature, is useful in time series analysis. Moreover, unlike DWT, MODWT is defined naturally for all sample sizes. The MODWT method was expanded while preserving the size of the data. There are various studies related to feature extraction [38]. Based on the dataset measured at Chosun University, a scalogram based on the continuous wavelet transform (CWT) is confirmed to be useful as an ECG feature [39]. By extending this, MODWT was selected as a feature suitable for the EMG analysis at edge devices. Owing to the MODWT method, feature extraction with light-weight can be realized. If the time series are directly analyzed, the size of the entire neural network for feature extraction increases, whereas in the MODWT method, a similar performance can be secured even with a small neural network. The basic implementation algorithm of the MODWT method is as follows. For a time series X with an arbitrary sample size N, the j-th level MODWT wavelet (W_j*) and scaling (V_j*) coefficients are defined as follows [40].

$$W_{j,t}^{*} = \sum_{l=0}^{L_j-1} h_{j,l}^{*} X_{t-l \ mod \ N} \tag{5}$$

$$V_{j,t}^{*} = \sum_{l=0}^{L_j-1} g_{j,l}^{*} X_{t-l \ mod \ N} \tag{6}$$

where $h_{j,l}{}^* = h_{j,l}/2^{j/2}$ are the MODWT wavelet filters, and $g_{j,l}{}^* = g_{j,l}/2^{j/2}$ are the MODWT scaling filters.

In this study, the second-order Daubechies filter (db2) was used, and the original signal was restored using inverse MODWT. In addition to db2, filters that can be used in the MODWT method include the first-order Daubechies (db1), fourth-order Daubechies (db4), Haar, Symlets, and Coiflets filters [41]. Except for the Haar and db1 filters, all exhibited excellent performance. Moreover, the decomposition level was selected as 4 because the performance increased up to level 4; however, the performance decreased by approximately 2% when the level was 5 or higher. This is related to the decomposition of the frequency domain. For example, the frequency spectrum of the level 1 MODWT signal is between 125 Hz and 250 Hz.

The results of the MODWT with the decomposition level of 4 are shown in Figure 4. Similar to the artificial intelligence network, the MODWT module will also be implemented inside the FPGA using digital logic. The results of each MODWT were provided as inputs to the artificial intelligence network.

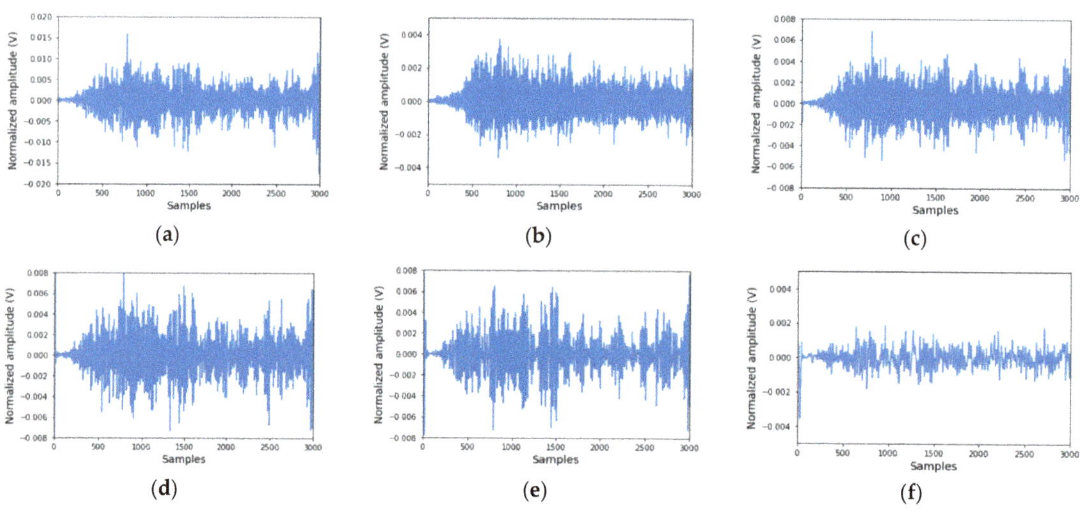

Figure 4. Maximal overlap discrete wavelet transform (MODWT) execution result. (**a**) Original signal, (**b**) level 1, (**c**) level 2, (**d**) level 3, (**e**) level 4, and (**f**) residual.

3. Artificial Intelligence Network

The MODWT signal for feature extraction has five channels and is given as an input to the one-dimensional convolution layer, as shown in Figure 5. The three one-dimensional convolution layers were used. The one-dimensional convolution layer is suitable for realizing a compressed neural network because the amount of computation is smaller than that of the two-dimensional convolution layer. To implement the light-weight characteristic while keeping the neural network as simple as possible, the accuracy was aimed at above 93%. Therefore, the verification of various neural networks was performed, and among them, a neural network with 93% or more accuracy was selected while achieving the light-weight characteristic. KerasTuner was used for the network design.

```
Model: "model"
_________________________________________________________________
 Layer (type)                Output Shape              Param #
=================================================================
 inputt_1 (InputLayer)       [(None, 3000, 5)]         0

 conv1dt_1 (Conv1D)          (None, 2998, 32)          480

 batcht_1 (BatchNormalizatio  (None, 2998, 32)         128
 n)

 relut_1 (Activation)        (None, 2998, 32)          0

 maxt_1 (MaxPooling1D)       (None, 999, 32)           0

 conv1dt_2 (Conv1D)          (None, 498, 16)           2560

 batcht_2 (BatchNormalizatio  (None, 498, 16)          64
 n)

 relut_2 (Activation)        (None, 498, 16)           0

 maxt_2 (MaxPooling1D)       (None, 166, 16)           0

 conv1dt_3 (Conv1D)          (None, 82, 16)            1024

 batcht_3 (BatchNormalizatio  (None, 82, 16)           64
 n)

 relut_3 (Activation)        (None, 82, 16)            0

 maxt_3 (MaxPooling1D)       (None, 20, 16)            0

 flatt_1 (Flatten)           (None, 320)               0

 denset_1 (Dense)            (None, 12)                3852

 denset_2 (Dense)            (None, 6)                 78

=================================================================
Total params: 8,250
Trainable params: 8,122
Non-trainable params: 128
_________________________________________________________________
```

Figure 5. Structure and parameters for an artificial intelligence network.

The activation function of the one-dimensional convolution layer used the rectified linear unit (ReLU), and padding was not used. The kernel for the convolution layer used L1 regulation, and the used parameter was 0.001. Three 1-dimensional convolution layers were used, and the number of channels was optimized. The max-pooling layer after the convolution layer was used to reduce the number of features. In the case of the fully connected layer (dense layer), the minimum features were used as input, and the sigmoid and softmax functions were used for the activation functions. These were implemented in the form of a look-up table when being implemented to the hardware resources in Section 4. Because the dense layer uses many parameters, the structure is implemented as concisely as possible. The focus in this case was to ensure high accuracy with few parameters and a simple structure.

In this case, the adaptive moment estimation (adam) optimizer was used for training, and the categorical cross-entropy function was used as the loss function. Moreover, the learning process was performed using the validation data of 30% of the total training data. The learning rate (*lr*), a representative hyperparameter, used a step decay function, and the *lr* change factor was set to 0.5. The initial *lr* value was set to 0.0001. The minimum value of *lr* was set to 10^{-7}. Thus, the appropriate learning rate was adjusted. The total number of trainable parameters of the designed neural network was 8122. This corresponds to the lightweight neural network, and 96% accuracy for the validation data was possible with the help of algorithms such as MODWT, data augmentation, and smoothing techniques. Thus, the implementation of a light-weight neural network that can be operated with a small number of resources, even when implemented as an edge device, was achieved. The accuracy and loss of training data and validation data according to the increase in training epochs are shown in Figure 6. The accuracies of the training data and validation data were approximately 99% and 96%, respectively.

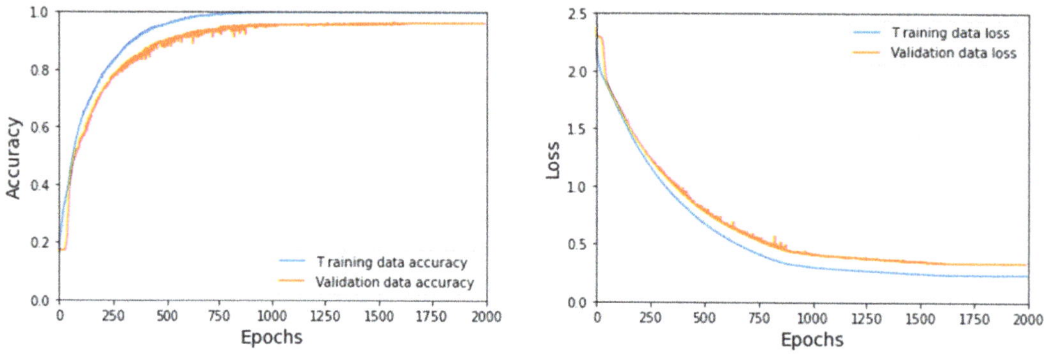

(**a**) (**b**)

Figure 6. (**a**) Accuracy of training and validation data. (**b**) Loss of training and validation data.

In the case of new data (test data), the confusion matrix is shown in Figure 7. The accuracy of the test data was 95% after training. Although this value has relatively low accuracy, the value is high considering that it is implemented using a small number of hardware resources. In the case of the other light-weight neural network example, the accuracy of "jet tagging" was approximately 74% [42]. Using MODWT, feature extraction is performed efficiently, and high accuracy can be secured with few parameters. In Figure 7, "spher" represents the action for holding spherical tools, "tip" for holding small tools, "palm" for grasping with the palm facing the object, "lat" for holding thin, flat objects, "cyl" for holding cylindrical tools, and "hook" for supporting a heavy load.

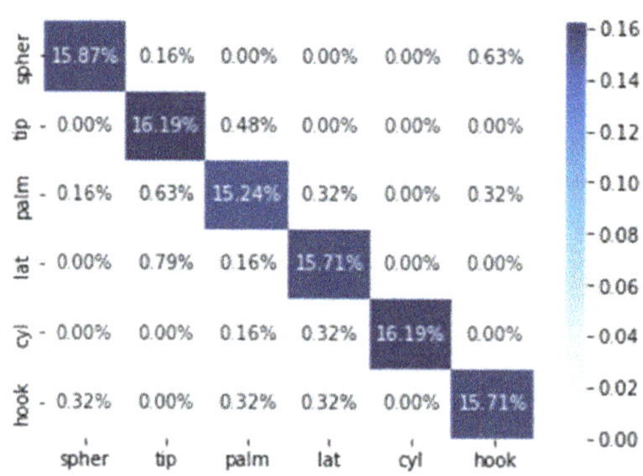

Figure 7. Confusion matrix for test data.

Finally, Figure 8 shows the receiver operating characteristic (ROC) curve and area under curve (AUC). The ROC curve is often used to evaluate the performance of a model that distinguishes classes, with a false positive rate (FPR) on the x-axis and a true positive rate (TPR) on the y-axis. Thus, various performance indices can be represented. Currently, the AUC is over 99% for the test data. In the previous study carried out by the authors of [43] using the State-of-the-Art (SOTA), the achievable accuracy with FPGAs was approximately 95.4% when using more parameters in the proposed network. In this study, higher accuracy was achieved with fewer parameters using various feature extraction techniques.

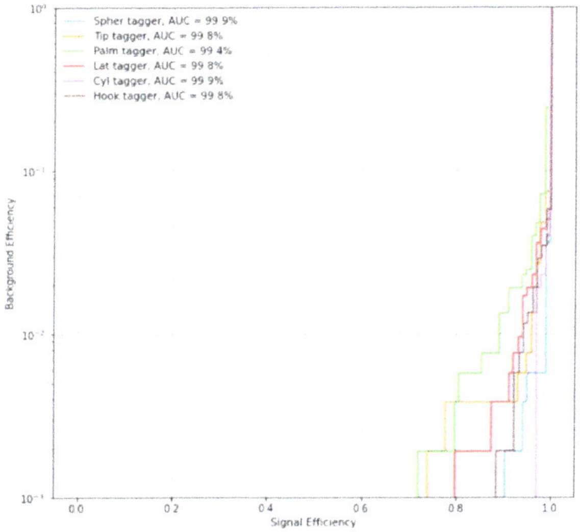

Figure 8. Receiver operating characteristic (ROC) curve for test data (six categories).

4. Hardware Deployment

Using the light-weight neural network model created by software using the Keras tool, it is converted to the Verilog hardware description language (HDL) using HLS [44]. Verilog HDL is changed to NAND or NOR gates through synthesis to create digital logic. HLS is a method that automatically converts the software code to Verilog HDL when written in C/C++, with which the user is familiar. Because large logic requires longer duration when written directly in a language such as Verilog HDL, HLS can significantly reduce the hardware development time and increase user convenience. Moreover, if the user uses a command such as "#pragma HLS PIPELINE", which is a separate grammar in HLS, the parallel computation of the FPGA can be accomplished. Therefore, latency reduction in the FPGA can be obtained. Thus, the structure optimization is performed together in the HLS. Figure 9 shows the HLS code for one-dimensional convolution layer written in C/C++, representatively. This layer proposed by Keras is configured using C/C++ and converted to Verilog HDL using HLS. The main algorithm is the process of adding the convolution operation of the input and filter after storing the values of the bias in the buffer. Each batch normalization, max-pooling, activation, and dense layer was implemented using HLS to be similar to the neural network proposed by Keras. In the case of the sigmoid and softmax functions, the number of exponential calculations is large; therefore, it is implemented in the form of a look-up table. Similar to the abovementioned method, several hardware implementation methods are available, such as Vivado SDAccel and NVIDIA Deep Learning Accelerator (NVDLA). The core part has the advantage of directly converting a neural network written in Caffe or similar frameworks into a register transfer level (RTL). In this study, the artificial intelligence part for EMG signal classification in real life can be implemented in an edge device, which was designed to have light-weight parameters through neural network design. This resulted in a reduction in hardware resources, and additional resource reduction was performed through additional bit optimization. The designed digital logic can be easily changed to application-specific integrated circuits (ASICs).

```
void conv1d(DTYPE in[NUM_INCHAN][IN_ROWS],
            const DTYPE filt[NUM_OUTCHAN][NUM_INCHAN][KSIZE],
            const DTYPE bias[NUM_OUTCHAN],
            DTYPE out[NUM_OUTCHAN][OUT_ROWS]){
    OFM: for(int ofm=0; ofm<NUM_OUTCHAN; ofm+=2){
        ROW_CLR: for(int r=0; r<OUT_ROWS; r++){
#pragma HLS PIPELINE
                acc_buf_0[r]=bias[ofm]; acc_buf_1[r]=bias[ofm+1];}
        IFM: for(int ifm=0; ifm<NUM_INCHAN; ifm++){
            ROW: for(int r=0; r<OUT_ROWS; r++){
#pragma HLS PIPELINE
                    acc_0=0; acc_1=0;
                    K_ROW: for(int k_r=0; k_r<KSIZE; k_r+=STRIDE){
                            acc_0 += filt[ofm][ifm][k_r]*in[ifm][r + k_r];
                            acc_1 += filt[ofm+1][ifm][k_r]*in[ifm][r + k_r];
                    }
                    acc_buf_0[r]+= acc_0; acc_buf_1[r]+= acc_1;
                }
        }
        ROW_CPY:for(int r=0;r<OUT_ROWS;r++){
#pragma HLS PIPELINE
            out[ofm][r] = acc_buf_0[r];
            out[ofm+1][r] = acc_buf_1[r];
        }
    }
}
```

Figure 9. High level synthesis (HLS) code for one-dimensional convolution layer.

The advantage of hardware deployment is that the number of systems can be freely used. Because the hardware is designed by us, the number of systems used inside can be defined and used; thus, the hardware resources can be reduced [45]. In the case of using the floating point number used in software in Section 3, 95% accuracy can be obtained for the test data using the hardware resource. This is the same accuracy result as that of the software. However, the number of bits used for the resource reduction is reduced to make it suitable for wearable devices. In this case, bit optimization was performed. In the case of a fixed point number, an error occurs, unlike in Keras, and a large difference is shown depending on the number of bits used. In the FPGA, the size of the fixed point number is determined through the "ap_fixed" keyword, and ap_fixed<16, 6> is the basic configuration for the FPGA. In particular, the entire bit becomes 16 bits, the integer is 6 bits, including the sign bit, and 10 bits represent the value below the decimal point. In this case, if ap_fixed<24, 6> is used, it can be implemented with a small number of hardware resources without degradation in accuracy. However, when ap_fixed<22, 6> was used, 80% of the hardware resources were used compared to ap_fixed<24, 6>, which was regarded as the optimal structure. In this case, the difference in accuracy was approximately 1%. However, if the number of bits is further reduced and the hardware is configured as ap_fixed<20, 6>, the accuracy is reduced to 86%. The distribution of the weights used to optimize the bit was examined, as shown in Figure 10. The total number of bits used for the optimal structure was 22. Bit optimization can also be performed using QKeras. QKeras is a quantization extension of Keras.

The network was implemented using HLS with ap_fixed<22, 6>. The accuracy of the test data was approximately 94%, and the AUC could be secured by more than 99%. If the basic ap_fixed<16, 6> is used, the accuracy is reduced to approximately 18%, and the AUC becomes more than 45%. No change is observed in accuracy when using floating-point numbers or ap_fixed<24, 6>. Moreover, a reuse factor from 10 to 20 was used to implement a structure that reuses resources completely. Figure 11 shows the ROC curve in the digital logic implemented in the FPGA with apiece<22, 6> compared with the Keras result. The AUC decreases with a decrease in the number of bits; an AUC can be secured by more than 99%.

When configured in an actual FPGA, communication was performed using the AXI structure, and the generated IP of the artificial neural network was used. The FPGA chipset "xcku5p-ffvb676-1-i" was used, and the device utilization is shown in Figure 12. In the case of AUC, it is shown as 99% or more for all classifications. In this case, the total operating speed for 4000 data at the time of inference is approximately 2.47 s when using the CPU and approximately 480 ms when using the proposed FPGA. Therefore, the operating speed for inference is approximately five times that of the proposed FPGA. In the case of FPGA, because it is composed of a wearable system, the data transmission time to the server is not required; therefore, a fast response time is guaranteed. Because security can be secured in healthcare services, the application of edge devices using FPGAs is expected to be of great help.

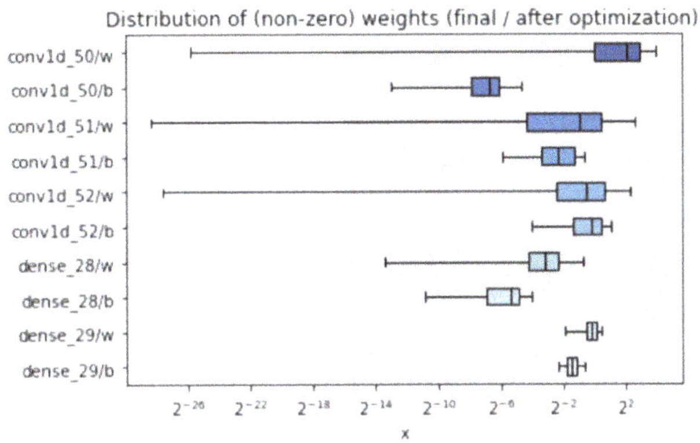

Figure 10. Bit selection using numerical profiling.

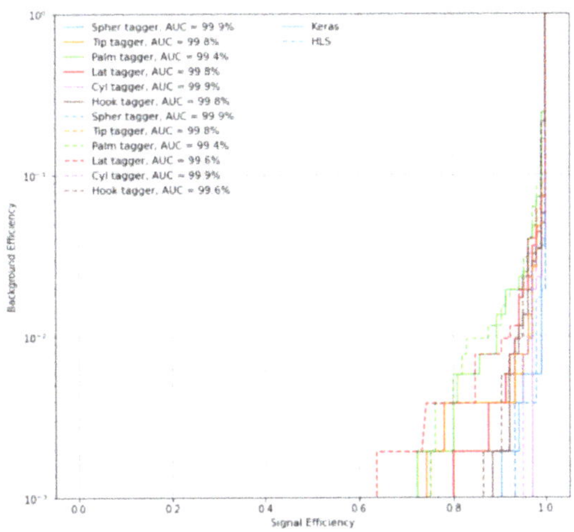

Figure 11. ROC curve for Keras and HLS (six categories).

Furthermore, fewer resources can be used by the pruning technique, although it is not implemented in the current structure. When the pruning technique is used with 50% sparsity, the accuracy is reduced by approximately 3%; hence, this option was not used.

Table 1 presents a comparison of the accuracy and response speed when performing inference using FPGA and CPU. The CPU was Intel Core i7-7700HQ. In the case of accuracy, test data were used. Thus, the EMG signal classification is possible using a small amount of resources when using an FPGA. Security is enhanced, and a fast response time can be acquired using FPGA. However, when using the HLS for hardware deployment, the performance of optimization between the latency and resources is rather insufficient; therefore, a more efficient implementation of the HLS should be considered in the future.

Name	BRAM_18K	DSP48E	FF	LUT	URAM
DSP	-	-	-	-	-
Expression	-	-	0	32	-
FIFO	480	-	11830	22236	-
Instance	163	498	67422	68379	-
Memory	-	-	-	-	-
Multiplexer	-	-	-	36	-
Register	-	-	6	-	-
Total	643	498	79258	90683	0
Available	960	1824	433920	216960	64
Utilization (%)	66	27	18	41	0

Figure 12. Device utilization for FPGA chipset "xcku5p-ffvb676-1-i".

Table 1. Comparison of accuracy for test data and response speed between CPU and FPGA.

Category	Accuracy	Inference Time (4000 Samples)
CPU	95%	2.47 s
FPGA, ap_fixed<24, 6>	95%	520 ms
FPGA, ap_fixed<22, 6>	94%	480 ms
FPGA, ap_fixed<20, 6>	86%	435 ms
FPGA, ap_fixed<16, 6>	18%	378 ms

EMG classification in real time has attracted considerable attention. In this regard, the response time was approximately 200 ms in the case of the MYO armband and 0.2 ms in the case of implementation with the MCU [46,47]. In this study, a response time of 0.12 ms per sample was obtained using the parallel computation and light weight of the FPGA (ap_fixed<22, 6>).

5. Conclusions and Discussion

The implementation of edge devices using EMG signals was studied. Among various edge devices, FPGAs were considered because they can easily perform parallel computation and can be implemented in ASICs in the future. In this case, owing to resource limitations in the FPGA, optimization was performed on the structure that could maintain accuracy while implementing the artificial neural network easily. For this, data augmentation was performed on the EMG signal, and MODWT, which is capable of time and frequency domain analysis, was used for the feature vector. In the case of convolutional and deep neural networks, the structure was optimized to prevent the number of parameters from exceeding 10,000, and the number of bits was optimized to maintain accuracy. Thus, HCI, disease diagnosis and user authentication could be performed quickly and with low power using artificial intelligence on a light-weight edge device. The implementation of wearable devices can contribute to security enhancement. The main contribution of this study is the examination of the practical applications of edge devices. However, to be grafted onto wearable devices, they must be implemented using fewer resources. In future studies, a wearable system with a bio-signal sensor system and edge device will be manufactured. This system will help the user's self-diagnosis at the desired time. Moreover, studies on Siamese or ensemble networks that can learn with less data are planned. Simultaneously, the study plans to manufacture PPG and ECG sensors to build a wearable system based on artificial intelligence.

Author Contributions: Conceptualization, H.-S.C.; methodology, H.-S.C.; software, H.-S.C.; validation, H.-S.C.; formal analysis, H.-S.C.; investigation, H.-S.C.; resources, H.-S.C.; data curation, H.-S.C.; writing—original draft preparation, H.-S.C.; writing—review and editing, H.-S.C.; visualization, H.-S.C.; supervision, H.-S.C.; project administration, H.-S.C.; funding acquisition, H.-S.C. All authors have read and agreed to the published version of the manuscript.

Funding: This research was supported by the Basic Science Research Program through the National Research Foundation of Korea (NRF) funded by the Ministry of Education (No. 2017R1A6A1A03015496).

Data Availability Statement: The data supporting the findings of the article are available in reference number [34].

Conflicts of Interest: The author declares no conflict of interest.

References

1. Swapna, M.; Viswanadhula, U.M.; Aluvalu, R.; Vardharajan, V.; Kotecha, K. Bio-Signals in Medical Applications and Challenges Using Artificial Intelligence. *J. Sens. Actuator Netw.* **2022**, *11*, 17. [CrossRef]
2. Hammad, M.; Pławiak, P.; Wang, K.; Acharya, U.R. ResNet-Attention model for human authentication using ECG signals. *Expert Syst.* **2021**, *38*, e12547. [CrossRef]
3. Choi, E.J.; Kim, D.K. Arousal and Valence Classification Model Based on Long Short-Term Memory and DEAP Data for Mental Healthcare Management. *Healthc. Inform. Res.* **2018**, *24*, 309–316. [CrossRef]
4. Shahid, H.; Butt, A.; Aziz, S.; Khan, M.U.; Naqvi, S.Z.H. Emotion Recognition System featuring a fusion of Electrocardiogram and Photoplethysmogram Features. In Proceedings of the 14th International Conference on Open Source Systems and Technologies, Lahore, Pakistan, 16–17 December 2020; pp. 1–6.
5. Yu, J.; Park, S.; Kwon, S.H.; Cho, K.H.; Lee, H. AI-Based Stroke Disease Prediction System Using ECG and PPG Bio-Signals. *IEEE Access* **2022**, *10*, 43623–43638. [CrossRef]
6. Muhammad, G.; Alshehri, F.; Karray, F.; El Saddik, A.; Alsulaiman, M.; Falk, T.H. A comprehensive survey on multimodal medical signals fusion for smart healthcare systems. *Inf. Fusion* **2021**, *76*, 355–375. [CrossRef]
7. Raurale, S.A.; McAllister, J.; Del Rincon, J.M. EMG biometric systems based on different wrist-hand movements. *IEEE Access* **2021**, *9*, 12256–12266. [CrossRef]
8. Rahim, M.A.; Shin, J. Hand movement activity-based character input system on a virtual keyboard. *Electronics* **2020**, *9*, 774. [CrossRef]
9. Antonelli, M.G.; Beomonte Zobel, P.; Durante, F.; Zeer, M. Modeling-Based EMG Signal (MBES) Classifier for Robotic Remote-Control Purposes. *Actuators* **2022**, *11*, 65. [CrossRef]
10. Mukhopadhyay, A.K.; Samui, S. An experimental study on upper limb position invariant EMG signal classification based on deep neural network. *Biomed. Signal Process. Control* **2020**, *55*, 101669. [CrossRef]
11. Albadawi, Y.; Takruri, M.; Awad, M. A review of recent developments in driver drowsiness detection systems. *Sensors* **2022**, *22*, 2069. [CrossRef] [PubMed]
12. Toro-Ossaba, A.; Jaramillo-Tigreros, J.; Tejada, J.C.; Pena, A.; Lopez-Gonzalez, A.; Castanho, R.A. LSTM Recurrent Neural Network for Hand Gesture Recognition Using EMG Signals. *Appl. Sci.* **2022**, *12*, 9700. [CrossRef]
13. Schluter, C.; Caraguay, W.; Ramos, D.C. Development of a low-cost EMG-data acquisition armband to control an above-elbow prosthesis. *J. Phys. Conf. Ser.* **2022**, *2232*, 012019. [CrossRef]
14. Alsolai, H.; Qureshi, S.; Zeeshan Iqbal, S.M.; Ameer, A.; Cheaha, D.; Henesey, L.E.; Karrila, S. Employing a Long-Short-Term Memory Neural Network to Improve Automatic Sleep Stage Classification of Pharmaco-EEG Profiles. *Appl. Sci.* **2022**, *12*, 5248. [CrossRef]
15. Sun, Y.; Xu, C.; Li, G.; Xu, W.; Kong, J.; Jiang, D.; Chen, D. Intelligent human computer interaction based on non-redundant EMG signal. *Alex. Eng. J.* **2020**, *59*, 1149–1157. [CrossRef]
16. Li, Q.D.P.; Zheng, J. Enhancing the security of pattern unlock with surface EMG-based biometrics. *Appl. Sci.* **2020**, *10*, 541. [CrossRef]
17. Sadikoglu, F.; Kavalcioglu, C.; Dagman, B. Electromyogram (EMG) signal detection, classification of EMG signals and diagnosis of neuropathy muscle disease. *Procedia Comput. Sci.* **2017**, *120*, 422–429. [CrossRef]
18. Ngai, W.K.; Xie, H.; Zou, D.; Chou, K.L. Emotion recognition based on convolutional neural networks and heterogeneous bio-signal data sources. *Inf. Fusion* **2022**, *77*, 107–117. [CrossRef]
19. Saikia, A.; Mazumdar, S.; Sahai, N.; Paul, S.; Bhatia, D. Performance analysis of artificial neural network for hand movement detection from EMG signals. *IETE J. Res.* **2022**, *68*, 1074–1083. [CrossRef]
20. Usman, M.; Amin, R.; Aldabbas, H.; Alouffi, B. Lightweight challenge-response authentication in SDN-based UAVs using elliptic curve cryptography. *Electronics* **2022**, *11*, 1026. [CrossRef]
21. Shumba, A.T.; Montanaro, T.; Sergi, I.; Fachechi, L.; De Vittorio, M.; Patrono, L. Leveraging IoT-Aware Technologies and AI Techniques for Real-Time Critical Healthcare Applications. *Sensors* **2022**, *22*, 7675. [CrossRef]
22. Zhu, G.; Liu, D.; Du, Y.; You, C.; Zhang, J.; Huang, K. Toward an intelligent edge: Wireless communication meets machine learning. *IEEE Commun. Mag.* **2020**, *58*, 19–25. [CrossRef]
23. Yazici, M.T.; Basurra, S.; Gaber, M.M. Edge machine learning: Enabling smart internet of things applications. *Big Data Cogn. Comput.* **2018**, *2*, 26. [CrossRef]
24. Sudharsan, B.; Breslin, J.G.; Ali, M.I. Edge2train: A framework to train machine learning models (SVMs) on resource-constrained IoT edge devices. In Proceedings of the 10th International Conference on the Internet of Things, Malmö, Sweden, 6–9 October 2020; pp. 1–8.

25. Akopyan, F.; Sawada, J.; Cassidy, A.; Alvarez-Icaza, R.; Arthur, J.; Merolla, P.; Modha, D.S. TrueNorth: Design and tool flow of a 65 mW 1 million neuron programmable neurosynaptic chip. *IEEE Trans. Comput. Aided Des. Integr. Circuits Syst.* **2015**, *34*, 1537–1557. [CrossRef]
26. Sambas, A.; Vaidyanathan, S.; Zhang, X.; Koyuncu, I.; Bonny, T.; Tuna, M.; Alcin, M.; Zhang, S.; Sulaiman, I.M.; Awwal, A.M.; et al. A Novel 3D Chaotic System With Line Equilibrium: Multistability, Integral Sliding Mode Control, Electronic Circuit, FPGA Implementation and Its Image Encryption. *IEEE Access* **2022**, *10*, 68057–68074. [CrossRef]
27. Sambas, A.; Vaidyanathan, S.; Bonny, T.; Zhang, S.; Hidayat, Y.; Gundara, G.; Mamat, M. Mathematical model and FPGA realization of a multi-stable chaotic dynamical system with a closed butterfly-like curve of equilibrium points. *Appl. Sci.* **2021**, *11*, 788. [CrossRef]
28. Sambas, A.; Vaidyanathan, S.; Tlelo-Cuautle, E.; Abd-El-Atty, B.; Abd El-Latif, A.A.; Guillen-Fernandez, O.; Sukono; Hidayat, Y.; Gundara, G. A 3-D multi-stable system with a peanut-shaped equilibrium curve: Circuit design, FPGA realization, and an application to image encryption. *IEEE Access* **2020**, *8*, 137116–137132. [CrossRef]
29. Liu, X.; Yang, J.; Zou, C.; Chen, Q.; Yan, X.; Chen, Y.; Cai, C. Collaborative edge computing with FPGA-based CNN accelerators for energy-efficient and time-aware face tracking system. *IEEE Trans. Comput. Soc. Syst.* **2021**, *9*, 252–266. [CrossRef]
30. Guo, K.; Zeng, S.; Yu, J.; Wang, Y.; Yang, H. A survey of FPGA-based neural network accelerator. *ACM Trans. Reconfigurable Technol. Syst.* **2019**, *12*, 1–26. [CrossRef]
31. Sandler, M.; Howard, A.; Zhu, M.; Zhmoginov, A.; Chen, L.C. Mobilenetv2: Inverted residuals and linear bottlenecks. In Proceedings of the IEEE Conference on Computer Vision and Pattern Recognition, Salt Lake City, UT, USA, 18–23 January 2018; pp. 4510–4520.
32. Norrie, T.; Patil, N.; Yoon, D.H.; Kurian, G.; Li, S.; Laudon, J.; Young, C.; Jouppi, N.; Patterson, D. The design process for Google's training chips: TPUv2 and TPUv3. *IEEE Micro* **2021**, *41*, 56–63. [CrossRef]
33. Han, S.; Mao, H.; Dally, W.J. Deep compression: Compressing deep neural networks with pruning, trained quantization and huffman coding. In Proceedings of the International Conference on Learning Representations, San Juan, Puerto Rico, 2–4 May 2016; pp. 1–14.
34. Sapsanis, C.; Georgoulas, G.; Tzes, A.; Lymberopoulos, D. Improving EMG based classification of basic hand movements using EMD. In Proceedings of the 35th Annual International Conference of the IEEE Engineering in Medicine and Biology Society 13 (EMBC 13), Osaka, Japan, 3–7 July 2013; pp. 5754–5757.
35. Lobov, S.; Krilova, N.; Kastalskiy, I.; Kazantsev, V.; Makarov, V.A. Latent factors limiting the performance of sEMG-interfaces. *Sensors* **2018**, *18*, 1122. [CrossRef]
36. Tsinganos, P.; Cornelis, B.; Cornelis, J.; Jansen, B.; Skodras, A. Data augmentation of surface electromyography for hand gesture recognition. *Sensors* **2020**, *20*, 4892. [CrossRef]
37. Jeong, J.W.; Lee, W.; Kim, Y.J. A Real-Time Wearable Physiological Monitoring System for Home-Based Healthcare Applications. *Sensors* **2021**, *22*, 104. [CrossRef]
38. Chen, Y.; Xia, R.; Zou, K.; Yang, K. FFTI: Image Inpainting Algorithm via Features Fusion and Two-Steps Inpainting. *J. Vis. Commun. Image Represent.* **2023**, *1*, 103776. [CrossRef]
39. Byeon, Y.H.; Pan, S.B.; Kwak, K.C. Intelligent deep models based on scalograms of electrocardiogram signals for biometrics. *Sensors* **2019**, *19*, 935. [CrossRef]
40. Adib, A.; Zaerpour, A.; Lotfirad, M. On the reliability of a novel MODWT-based hybrid ARIMA-artificial intelligence approach to forecast daily snow depth (Case study: The western part of the Rocky Mountains in the USA). *Cold Reg. Sci. Technol.* **2021**, *189*, 103342. [CrossRef]
41. Zhdanov, D.S.; Zemlyakov, I.Y.; Kosteley, Y.V.; Bureev, A.S. Choice of Wavelet Filtering Parameters for Processing Fetal Phonocardiograms with High Noise Level. *Biomed. Eng.* **2021**, *55*, 194–199. [CrossRef]
42. Fast Machine Learning Lab. Available online: https://github.com/fastmachinelearning/ (accessed on 1 February 2022).
43. Kang, S.; Kim, H.; Park, C.; Sim, Y.; Lee, S.; Jung, Y. sEMG-Based Hand Gesture Recognition Using Binarized Neural Network. *Sensors* **2023**, *23*, 1436. [CrossRef]
44. Westby, I.; Yang, X.; Liu, T.; Xu, H. FPGA acceleration on a multi-layer perceptron neural network for digit recognition. *J. Supercomput.* **2021**, *77*, 14356–14373. [CrossRef]
45. Xia, M.; Huang, Z.; Tian, L.; Wang, H.; Chang, V.; Zhu, Y.; Feng, S. SparkNoC: An energy-efficiency FPGA-based accelerator using optimized lightweight CNN for edge computing. *J. Syst. Archit.* **2021**, *115*, 101991. [CrossRef]
46. Zhang, Z.; Yang, K.; Qian, J.; Zhang, L. Real-time surface EMG pattern recognition for hand gestures based on an artificial neural network. *Sensors* **2019**, *19*, 3170. [CrossRef]
47. Liu, X.; Sacks, J.; Zhang, M.; Richardson, A.G.; Lucas, T.H.; Van der Spiegel, J. The virtual trackpad: An electromyography-based, wireless, real-time, low-power, embedded hand-gesture-recognition system using an event-driven artificial neural network. *IEEE Trans. Circuits Syst. II Express Briefs* **2016**, *64*, 1257–1261. [CrossRef]

Disclaimer/Publisher's Note: The statements, opinions and data contained in all publications are solely those of the individual author(s) and contributor(s) and not of MDPI and/or the editor(s). MDPI and/or the editor(s) disclaim responsibility for any injury to people or property resulting from any ideas, methods, instructions or products referred to in the content.

electronics

Article

FPGA Implementation of Shack–Hartmann Wavefront Sensing Using Stream-Based Center of Gravity Method for Centroid Estimation

Fanpeng Kong [1], Manuel Cegarra Polo [2] and Andrew Lambert [1,*]

[1] School of Engineering and Information Technology, University of New South Wales, Canberra, ACT 2612, Australia
[2] Japan Aerospace Exploration Agency, Tokyo 182-8522, Japan
* Correspondence: a-lambert@adfa.edu.au

Abstract: We present a fast and reconfigurable architecture for Shack–Hartmann wavefront sensing implemented on FPGA devices using a stream-based center of gravity to measure the spot displacements. By calculating the center of gravity around each incoming pixel with an optimal window matching the spot size, the common trade-off between noise and bias errors and dynamic range due to window size existing in conventional center of gravity methods is avoided. In addition, the accuracy of centroid estimation is not compromised when the spot moves to or even crosses the sub-aperture boundary, leading to an increased dynamic range. The calculation of the centroid begins while the pixel values are read from an image sensor and further computation such as slope and partial wavefront reconstruction follows immediately as the sub-aperture centroids are ready. The result is a real-time wavefront sensing system with very low latency and high measurement accuracy feasible for targeting on low-cost FPGA devices. This architecture provides a promising solution which can cope with multiple target objects and work in moderate scintillation.

Keywords: adaptive optics (AO); Shack–Hartmann wavefront sensor (SHWFS); wavefront sensing; field-programmable gate array (FPGA)

Citation: Kong, F.; Cegarra Polo, M.; Lambert, A. FPGA Implementation of Shack–Hartmann Wavefront Sensing Using Stream-Based Center of Gravity Method for Centroid Estimation. *Electronics* **2023**, *12*, 1714. https://doi.org/10.3390/electronics12071714

Academic Editors: Andres Upegui, Andrea Guerrieri and Laurent Gantel

Received: 27 February 2023
Revised: 27 March 2023
Accepted: 28 March 2023
Published: 4 April 2023

Copyright: © 2023 by the authors. Licensee MDPI, Basel, Switzerland. This article is an open access article distributed under the terms and conditions of the Creative Commons Attribution (CC BY) license (https://creativecommons.org/licenses/by/4.0/).

1. Introduction

The Shack–Hartmann wavefront sensor (SHWFS) is one of the most widely used wavefront sensors (WFSs) in adaptive optics (AO) systems [1]. The micro lenslet array (MLA) samples the input wavefront spatially, and by determining the local slope on each sub-aperture, the entire wavefront can be reconstructed successively. The local slopes over individual sub-apertures are calculated linearly from the displacements of spots focused by the micro lenses from their optical axes. Therefore, the accuracy of the spot displacement estimation is directly related to the overall performance of the SHWFS, as error from the estimation will pass through various stages to the final wavefront reconstruction stage. Center of Gravity (CoG)-based methods have been largely used for determining the centroids of the spots by measuring the center of mass within a window associated to a particular micro lenslet. Matched filter [2], minimum mean-square-error estimator [3], and maximum-likelihood methods [4] have been studied to estimate the centroid or wavefront slope as well. If the source is an extended object instead of a point source, cross-correlation methods can also be used to estimate the sub-image displacement [5–7]. A comparison of some commonly used centroiding algorithms, including thresholding (TCoG), weighted centroid (WCoG), correlation, and quad cell, is given by Thomas et al. [8]. For closed-loop AO systems and fast wavefront sensing, the CoG-based methods are still preferred over other methods due to their robustness and easy implementation.

Conventional CoG estimations are often corrupted due to various noise sources such as photon noise, readout noise of the image sensor, and the finite and coarse sampling of the pixels. These CoG-derived methods also fail to estimate the local wavefront if the spot breaks into parts due to strong turbulence with lower Fried coherence length r_0 or when scintillation from distant volumes of turbulence is presented. Using Gaussian approximation, Rousset [9] pointed out that the noise variance in local wavefront estimation due to sensor readout noise increases with the increased number of pixels for CoG calculation. On the other hand, Irwan et al. showed that the error variance due to photon noise also diverges as detector size increases, even for a perfect CCD array, and even without readout noise and effects due to finite pixel size. These analyses mean that a small CoG calculation window is necessary in order to reduce the error contribution of photon and sensor readout noise. However, if the window for the CoG calculation is too small, signal truncation error will be introduced when the spot moves to the window edge, which in turn leads to a limited dynamic range of the SHWFS. Therefore, the optimal CoG calculation window size often needs to balance between the noise errors and dynamic range. To isolate the useful signal for CoG calculation, the traditional CoG methods have been improved by thresholding the signal [10,11] or adding weight to emphasize the signal [12,13]. The error sources for centroid computation of a point source on a CCD-based sensor were analyzed by Ma et al. [14] and the best threshold level is given. Other methods using iterative detection of spot location and centroid estimation have been reported [15,16]. While they improve the best area for the CoG calculation, these iterations will introduce extra delay for the centroid measurements and eventually reduce the bandwidth of a closed-loop AO system. Recent efforts to improve the performance of Shack–Hartmann WFS include direct wavefront reconstruction as a continuous function from a bitmap image of the Shack–Hartmann pattern [17], and using artificial neural networks [18] and learning-based methods [19] for spot detection. To overcome the limitation of Shack–Hartmann WFS in certain situations, e.g., under strong scintillation, a diffractive lenslet array can also been used to replace the physical MLA, leading to a more flexible and adaptable Shack–Hartamnn WFS [20]. Talmi and Ribak [21] showed that gradient calculation over the whole aperture is possible by direct demodulation on the grid without reverting to Fourier Transforms. This method is especially suited to very large arrays due to the saving of computation by removing the two inverse Fourier Transforms. Importantly, they considered that incomplete spots, for example, at the edge of the aperture, would create bias on the complete reconstruction, and showed that processing these in a sensible way in the image domain could have a lesser effect on the whole reconstruction.

In addition to the effort to improve the accuracy of centroid estimation algorithms, other researchers also tried to increase the wavefront sensing speed by utilizing special hardware such as GPU [22,23] or field-programmable gate array (FPGA) devices for implementation. For example, FPGA devices have been used both in complex AO systems to process data where the timing is crucial [24–26], or used to implement centroid estimation and reconstruction [27–29], or even to develop full AO application including driving the wavefront corrector [30,31]. In comparison with conventional CPU [32] or GPU-based solutions, FPGA devices provide a cost-effective way to achieve a high throughput, low latency and reconfigurable wavefront sensing, and AO system thanks to their parallel computation power.

In our previous work [33], we proposed an improved stream-based center of gravity (SCoG) method for centroid estimation which is suitable to be implemented on FPGA devices. By extending the conventional CoG method to evaluate the center of gravity around each incoming pixel, the SCoG method can use an optimal CoG window matching the size of the spot behind the MLA without the common trade-off between increased bias error and reduced noise errors. In addition, the accuracy of the centroid estimation by SCoG is not compromised when the spot moves to the sub-aperture edge or even crosses the boundary, since the CoG operation centers on each individual pixel. The SCoG is also able to detect multiple centroids within one sub-aperture when the size of the CoG window is chosen appropriately because of its whole sensor centroid calculation.

While complicated and advanced CoG methods [12,13,15,17] have been proposed to improve the accuracy of the centroid estimation, little work has been reported to discuss their appropriate implementations to meet the low latency and high bandwidth requirement in real-time AO systems. However, most real-time implementations of the Shack–Hartmann WFS [22,23,28,32] use the basic thresholding CoG method. In this paper, we present a complete Shack–Hartmann wavefront sensing system implemented on FPGA hardware with very low latency and high accuracy using the superior SCoG for centroid estimations. A parallel slope calculation and a robust least-squares wavefront reconstruction are also implemented in a pipe-lined way after the centroid estimation. The paper is organized as follows: In Section 2, the theory of stream-based center of gravity deriving from conventional CoG methods, special treatments of multiple or missing centroids in sub-apertures, and the modal wavefront reconstruction are explained. The hardware implementations of the SCoG module, centroids segmentation module, and least-square modal wavefront reconstruction module are described in detail in Section 3. In Section 4, the resource usage and latency of the FPGA implementation are analyzed. Performance of the centroiding algorithm is compared with a traditional CoG method using an artificially generated image of spots followed by an examination of the wavefront reconstruction performance of the whole Shack–Hartmann WFS system. Conclusions and future work are summarized in Section 5.

2. Theoretical Background

Some notations used in this section to describe the stream-based center of gravity algorithm are listed in Table 1.

Table 1. Notations.

Symbol	Description
(i,j)	pixel indices
(r,c)	sub-aperture indices
(p,q)	stream centroids indices
$\mathcal{A}(r,c)$	sub-aperture
$\mathcal{C}(p,q)$	stream centroid
$\hat{C}_x(r,c), \hat{C}_y(r,c)$	centroid estimation in $\mathcal{A}(r,c)$
$s_x(r,c), s_y(r,c)$	average slope in $\mathcal{A}(r,c)$
$I(x_i,y_j)$	image intensity at pixel (x_i,y_j)

2.1. Conventional Center of Gravity

The geometric diagram of a single lenslet from a MLA is shown in Figure 1. The local wavefront tilt θ in a sub-aperture $\mathcal{A}(r,c)$ causes a shift Δx of the focal spot from its reference on-axis position $(C_{x,ref}(r,c), C_{y,ref}(r,c))$ when a plane wavefront is used. The size of the diffraction-limited spot is determined by the f-number of the lenslet and equals to $2.44\lambda f_{ML}/D_{ML}$ where f_{ML} and D_{ML} are the focal length and diameter of the micro lens respectively. By determining the local slopes at all sub-apertures, a continuous wavefront map can be reconstructed using either zonal- or modal-based reconstruction methods.

The local slope over sub-aperture $\mathcal{A}(r,c)$ can be calculated by:

$$s_x(r,c) = \frac{\Delta_x(r,c)}{f_{ML}} = \frac{\hat{C}_x(r,c) - C_{x,ref}(r,c)}{f_{ML}} \tag{1a}$$

$$s_y(r,c) = \frac{\Delta_y(r,c)}{f_{ML}} = \frac{\hat{C}_y(r,c) - C_{y,ref}(r,c)}{f_{ML}} \tag{1b}$$

where $\Delta_x(r,c)$, $\Delta_y(r,c)$ are the displacement of the spot in x and y directions from its reference location.

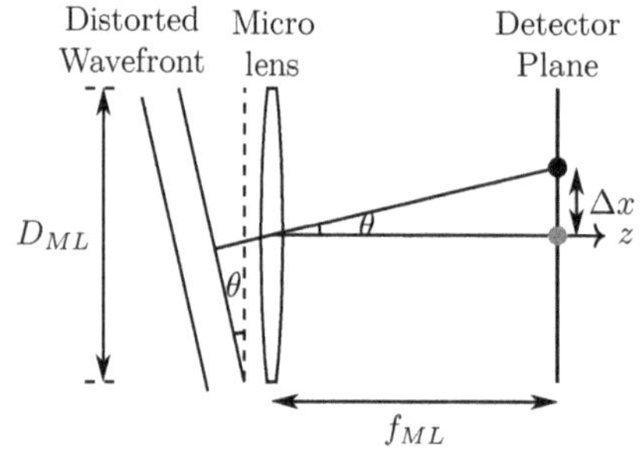

Figure 1. Schematic of a single micro lens in the Shack–Hartmann wavefront sensor.

In Equation (2), the centroid in the sub-aperture $\mathcal{A}(r,c)$ can be measured in x and y directions using the traditional CoG definition as:

$$\hat{C}_x(r,c) = \frac{\sum_{x_i}\sum_{y_j} x_i I(x_i,y_j)}{\sum_{x_i}\sum_{y_j} I(x_i,y_j)} \tag{2a}$$

$$\hat{C}_y(r,c) = \frac{\sum_{x_i}\sum_{y_j} y_j I(x_i,y_j)}{\sum_{x_i}\sum_{y_j} I(x_i,y_j)} \tag{2b}$$

where r,c is the coordinate of the sub-aperture, x_i, y_j is pixel index, and $I(x_i,y_j)$ is the pixel intensity. In a traditional instrument, the CoG is only calculated on the central pixel location $i = kr, j = kc$ per lenslet.

2.2. Stream-Based Center of Gravity

The conventional CoG method can be extended by evaluating the centroid estimation on each pixel of the signal:

$$\hat{C}_x(i,j) = \frac{\sum_{m=-M}^{M}\sum_{n=-N}^{N} F_x(m,n) I(x_i,y_j)}{\sum_{m=-M}^{M}\sum_{n=-N}^{N} I(x_i,y_j)} \tag{3a}$$

$$\hat{C}_y(i,j) = \frac{\sum_{m=-M}^{M}\sum_{n=-N}^{N} F_y(m,n) I(x_i,y_j)}{\sum_{m=-M}^{M}\sum_{n=-N}^{N} I(x_i,y_j)} \tag{3b}$$

where $F(m,n)$ is a linear filter ranging from $-M$ to M. $\hat{C}_x(i,j), \hat{C}_y(i,j)$ represents the estimated centroid value for the pixel (i,j) within a square window of side size of $2M+1$.

$$F_x(m,n) = \begin{pmatrix} -M & -M+1 & \cdots & M \\ -M & -M+1 & \cdots & M \\ \vdots & \vdots & \ddots & \vdots \\ -M & -M+1 & \cdots & M \end{pmatrix} \tag{4a}$$

$$F_y(m,n) = \begin{pmatrix} -N & -N & \cdots & -N \\ -N+1 & -N+1 & \cdots & -N+1 \\ \vdots & \vdots & \ddots & \vdots \\ N & N & \cdots & N \end{pmatrix} \tag{4b}$$

If the centroid estimation value at a pixel (i_p, j_q) equals to zero, then a spot is centered on this pixel. In most cases, however, the centroid is less likely to sit on an exact pixel but rather between two pixels. A potential spot is detected around pixel (i_p, j_q) if a zero-crossing (from positive to negative) of CoG values happens horizontally between pixels $(i_p - 1, j_q)$ and (i_p, j_q) and vertically between pixels $(i_p, j_q - 1)$ and (i_p, j_q) at the same time.

The sub-pixel shift in the x directions from pixel (i_p, j_q) can be interpreted linearly by the CoG values at pixel (i_p, j_q) and its left pixel $(i_p - 1, j_q)$, while the sub-pixel shift in y direction can be interpreted similarly by the CoG values at pixel (i_p, j_q) and its above pixel $(i_p, j_q - 1)$:

$$\Delta_x(i_p, j_q) = \frac{\hat{C}_x(i_p, j_q)}{\hat{C}_x(i_p - 1, j_q) - \hat{C}_x(i_p, j_q)} \quad (5a)$$

$$\Delta_y(i_p, j_q) = \frac{\hat{C}_y(i_p, j_q)}{\hat{C}_y(i_p, j_q - 1) - \hat{C}_y(i_p, j_q)} \quad (5b)$$

Therefore, the (p, q)th centroid $\mathcal{C}(p, q)$ where a potential spot is located can be described by $(\hat{x}_p, \hat{y}_q)$ as below:

$$\hat{x}_p = i_p + \Delta_x(i_p, j_q) \quad (6a)$$

$$\hat{y}_q = j_q + \Delta_y(i_p, j_q) \quad (6b)$$

Note that the calculation of integer and decimal parts of centroid $\mathcal{C}(p, q)$ are through separate steps and the integer parts, i.e., pixel indices, are determined first. (i_p, j_q) can be obtained directly by the sign changes of the numerators in Equation (3) as the denominators which represent the sum of energy within the kernel window are always positive. If only whole pixel resolution is concerned, calculations of the numerators are sufficient to locate the pixel indices.

In this work, an estimate of centroid is made based on each and every pixel streamed from the image sensor, for which best centroids are tagged resulting in a stream of centroids or SCoG synchronous with the stream of pixels. Several immediate advantages of SCoG over conventional CoG-based centroid estimation methods can be noticed [33]. Since the CoG window is floating with the incoming pixels and will center around each potential genuine centroid, bias errors due to asymmetric CoG filter are largely avoided. In addition, the size of the CoG window $2M + 1$ can be optimised by matching with the diffraction-limited spot size to minimize the influence of irrelevant pixels so that noise errors are minimised.

It is worth noting the unique characteristics of the centroids detected by the stream-based CoG algorithm. Using conventional sub-aperture-based CoG methods, only one centroid will be estimated for each sub-aperture, even when multiple spots exist due to, for example, binary star structure or strong turbulence, resulting in a "broken" spot (r_0 is less than the microlens diameter). In addition, the measured centroids belonging to each sub-aperture are rather apparent from conventional CoG methods. However, the stream of centroids from SCoG arises in order based on the position of the spot occurrence in the input image frame as it is read from the image sensor, row-by-row, and column-by-column. Therefore, the SCoG centroids stream need to be further processed in order to be used in the conventional zonal or modal wavefront reconstruction algorithms.

2.3. *Segmentation of the Streamed Centroids*

There are two problems that need to be addressed for the stream of centroids $\mathcal{C}(p, q)$ in order to get the centroid estimation for each sub-aperture associated with traditional CoG-based methods. First, the occurrence of centroid $\mathcal{C}(p, q)$ follows the lower row number to a higher row number or a lower column number to a higher column number depending on the pixel reading sequence of the particular sensor. For a particular centroid $\mathcal{C}(p, q)$, it needs to be assigned to a sub-aperture $\mathcal{A}(r, c)$ if its values are within the window of $\mathcal{A}(r, c)$ defined by:

$$r = \left\lfloor \frac{i_p}{w} \right\rfloor, c = \left\lfloor \frac{j_q}{w} \right\rfloor \quad (7)$$

where $\lfloor \cdot \rfloor$ is the floor operation and w is the window width in pixels.

Secondly, it is possible that multiple centroids or no valid centroid are detected within one sub-aperture $\mathcal{A}(r, c)$, perhaps because of multiple objects, obstruction, or scintillation.

For the multiple centroids case, different strategies can be used, such as using the centroid with the highest energy (sum of intensity as the denominator in Equation (3)) or taking the average of all centroids.

In our current implementation, we used the average of all centroids in the sub-aperture, hence effectively measuring the G-tilt of the local wavefront:

$$\left.\begin{aligned}\hat{C}_x(r,c) &= \langle \hat{x}_p \rangle \\ \hat{C}_y(r,c) &= \langle \hat{y}_q \rangle\end{aligned}\right\} \quad \text{where}(i_p, j_q) \in \mathcal{A}(r,c) \tag{8}$$

where $\langle \cdot \rangle$ denotes the average operation and $\in$ means the integer part of $\mathcal{C}(p,q)$ that falls within the pixel range of sub-aperture $\mathcal{A}(r,c)$.

On the other hand, if a centroid is missing in a sub-aperture $\mathcal{A}(r,c)$, it is possible to generate an average one from its surrounding sub-apertures. However, if one or more surrounding sub-apertures also miss valid centroids, the chain of average operation will expand to further sub-apertures which could soon become too complicated to manage. The other method to treat the sub-aperture with missing centroid is to inherit the corresponding centroid from a previous frame, which could be traced back to an original reference centroid if none of the previous frames contain a valid centroid.

Once all the valid sub-apertures in one MLA row have been processed to select a representative centroid estimation, they need to be re-ordered to match the sequence of the physical geometry, so the following wavefront reconstruction can be started even though the remaining lenslets have not yet arrived. Therefore, the further processing of the streamed centroids follows a *sorting–processing–reordering* procedure.

2.4. Wavefront Reconstruction

From the discrete slopes at all the valid sub-apertures as given by Equation (1), a continuous wavefront map can be reconstructed using *zonal*, *modal*, or *FFT-based* methods [34,35]. Using the modal wavefront reconstruction, the incoming wavefront $W(x,y)$ over the pupil can be decomposed by a set of orthogonal functions, such as Zernike polynomials:

$$W(x,y) = \sum_{k=1}^{N} a_k Z_k(x,y) \tag{9}$$

where a_k represents the weight of the kth Zernike term $Z_k(x,y)$ and N is the total number of Zernike modes used to approximate the actual wavefront.

In Equation (1), the local slope of wavefront on the individual sub-apertures is expressed as a relation between the local sub-image displacements for each axis $(\Delta_x(r,c), \Delta_y(r,c))$ and the MLA focal length f_{ML}. Considering only the x axis results in the following equation:

$$s_x(r,c) = \left.\frac{\partial W(x,y)}{\partial x}\right|_{(r,c)} = \frac{\Delta x(r,c)}{f_{ML}} \tag{10}$$

By combining Equations (9) and (10), the gradients of the wavefront over each sub-aperture can be related with a weighted sum of Zernike polynomials as follows:

$$\frac{\Delta x(r,c)}{f_{ML}} = \sum_{k=1}^{N} a_k \left.\frac{\partial Z_k(x,y)}{\partial x}\right|_{(r,c)} \tag{11}$$

Considering the slopes in both x and y directions for all the sub-apertures, Equation (11) can be expressed in the following matrix form:

$$\boldsymbol{s}_{2M\times 1} = \boldsymbol{W}_{2M\times N}\boldsymbol{a}_{N\times 1} \tag{12}$$

where M is the total number of valid sub-apertures and N is the number of Zernike modes used for wavefront reconstruction. $\boldsymbol{s}$ is the slope vector of dimensions $2M \times 1$ and $\boldsymbol{a}$ is the

Zernike mode coefficients vector of dimension $N \times 1$. $\boldsymbol{W}$ is a matrix of dimension $2M \times N$ whose elements are the partial derivative of a Zernike mode on either x or y direction. Equation (12) defines $2M$ linear equations. To obtain the Zernike coefficients vector $\boldsymbol{a}$ from measured slope vector $\boldsymbol{s}$, the pseudo-inverse of matrix $\boldsymbol{W}$ is used:

$$\boldsymbol{a}_{N\times 1} = \boldsymbol{E}_{N\times 2M}\boldsymbol{s}_{2M\times 1}; \tag{13}$$

The pseudo-inverse matrix $\boldsymbol{E}$, also known as the calibration matrix, has a dimension of $N \times 2M$ and can be calculated using the least squares estimation method:

$$\boldsymbol{E} = (\boldsymbol{W}^T\boldsymbol{W})^{-1}\boldsymbol{W}^T \tag{14}$$

In Section 3.4, the detailed implementation of the modal wavefront reconstruction is explained.

3. Implementation

In this section, a parallel implementation of the stream-based center of gravity algorithm and the modal wavefront reconstruction suitable for FPGA devices are described. By taking advantage of parallelism and storage resources of FPGA devices, the CoG operation can be evaluated at each incoming pixel in real-time. The wavefront reconstruction can also start partially as soon as the centroids estimation of sub-apertures becomes available and complete at a very short delay after acquiring one image frame (and certainly before the start of the next frame).

The overall block diagram of the SHWFS system design using the stream-based CoG method is shown in Figure 2. The complete implementation consists of four main modules corresponding to Sections 2.2–2.4: *SCoG* module computes centroid on all incoming pixels and presents a stream of valid centroids; *SEG* module sorts the centroids to confined sub-apertures and handles multiple centroids or centroids missing in some sub-apertures; *SLOPE* module calculate the local slopes from measured centroids and can also be used to generate a reference centroid grid; *RECON* module conducts a modal wavefront reconstruction from the measured slopes. The SCoG and SEG modules together provide similar functions to other conventional CoG methods. The auxiliary pre-processing, display, and control modules shown in light blue blocks are necessary to configure the image sensor, set system parameters, and visualize various results but not directly related to the research interest and therefore are omitted in the following discussion.

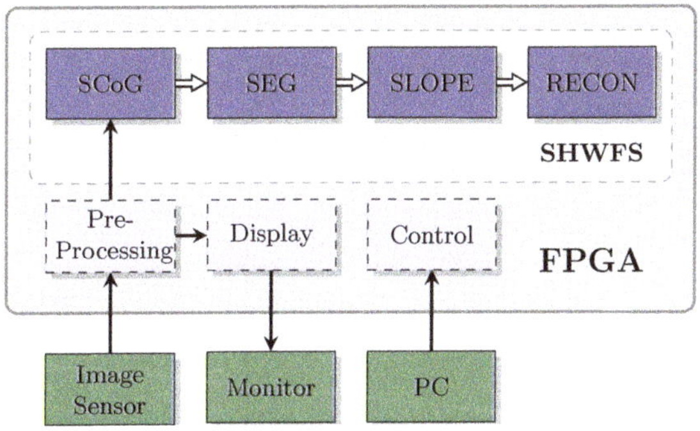

Figure 2. Top level block diagram of the SHWFS implementation where the implementation of the SHWFS is described here.

3.1. SCoG Module

In Figure 3, the 2D pixel array of an image sensor at the focal plane of the MLA is shown. The sub-aperture window SA corresponds to individual lenslet with its size and number of pixels (5 × 5 for this example) determined by the lenslet and pixel sizes.

Traditional CoG-based methods use all the pixels (except for the windowing CoG) inside the SA window for the calculation and generally need to read the whole image frame before computation. The SCoG operates on a much smaller pixel set (the blue dashed window) which could be optimized by matching the window with the spot size (3 × 3 for this example). In the FPGA devices, First-In, First-Out (FIFO), or Look Up Table (LUT) can be used to buffer several rows and columns of pixels, so as the last required pixel arrives (light green and light blue pixel for the first and last possible centroids in SA_{11}), the computation of the CoG will start.

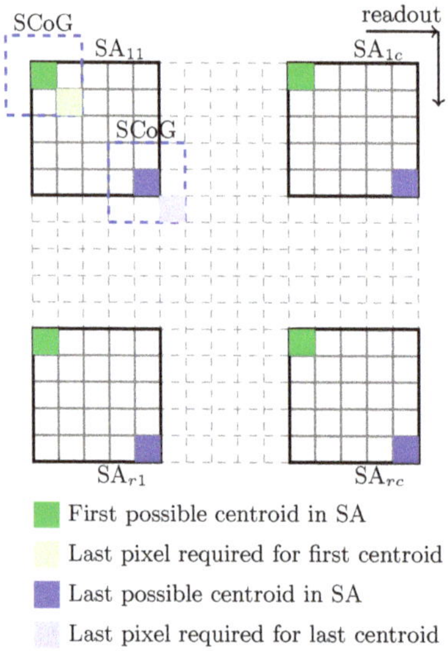

Figure 3. Illustration of the sub-aperture and SCoG window on the image sensor pixel array.

A high level logic diagram for calculating the stream centroid according to Section 2.2 on each incoming pixel is shown in Figure 4, where pixel Data D_{IN} is synchronous with the master clock. The circuit in the yellow box represents a 2D convolution operation and it calculates either the numerators or denominators in Equation (3) depending on the chosen coefficients C_{11} to C_{33} in the multipliers, which represent the matrix elements in Equation (4). The SCoG window is assumed with a 3 × 3 size, and therefore three buffers are used to store three rows of pixel values and three Flip-Flop (FF) registers are used to buffer three columns of pixels. As the last required pixel within the SCoG window arrives at the output of the first buffer, all the required pixel values will appear at the nine registers' output simultaneously at the next clock cycle. This implementation makes logical sense but is impractical, as all the multiplications and additions need to be finished in one clock cycle. This causes challenges to meeting timing constraints in FPGA implementation. In Appendix A, matrix multiplication is separated to two 1D filter operations, which reduces the usage of multipliers. Furthermore, the implementation of the two 1D filter operations are fully pipe-lined to best improve the timing closure in trade of more latency.

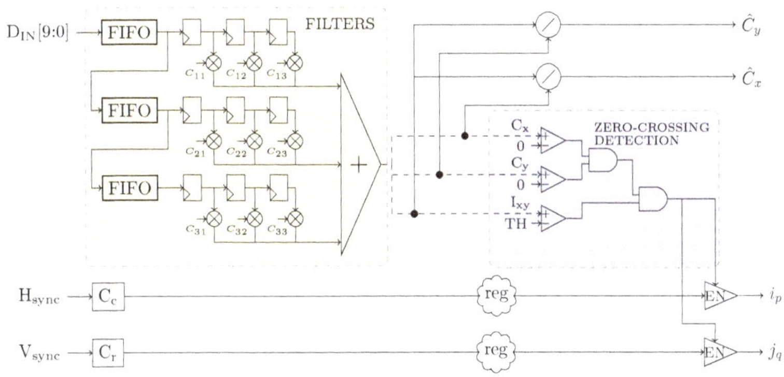

Figure 4. Diagram for the implementation of SCoG module used for calculate integer pixel centroid values. Depending on the coefficient values **C**, the numerators and denominators in Equation (3) are calculated from the FILTERS circuit.

The zero-crossing detection circuit is shown in the blue box where the numerators and denominators in Equation (3) are present simultaneously at the input. The horizontal and vertical zero-crossing can be determined by checking the sign of the numerators while the sum of the intensity, i.e., the denominator, can be compared with a threshold value I_{TH} to eliminate fake centroids due to noise pixels. The two dividers calculate the centroid at pixel (i_p, j_q) as described by Equation (3). The registers on the H_sync and V_sync lines are necessary to introduce the same number of clock delays for the counters of row and column numbers as those FILTERS, and ZERO_CROSSING DETECTION circuits. The length of these is a direct measure of the latency of the calculation.

The implementation of the sub-pixel interpolation in the x and y directions according to Equations (5) and (6) is shown in Figure 5. To interpret the sub-pixel shift $\Delta_x(i_p, j_q)$, the centroid estimation from the previous pixel is required, which can be buffered by a FF register. For the sub-pixel shift in the y direction, however, the pixel in the previous row is needed, which means the whole row of centroids value in y direction needed to be buffered in a FIFO with depth that is the same as the column number.

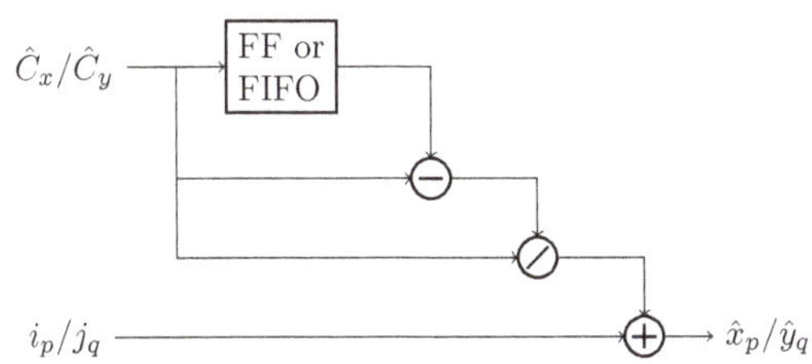

Figure 5. Diagram for the implementation sub-pixel interpolation in x and y direction.

3.2. Segmentation Module

The stream-based CoG algorithm itself is able to determine multiple spots presented in a sub-aperture if the filter size is not so large as to cover all the spots. Recall that the filter size is matched to the spot size, not the sub-aperture size Equation (3). How to use these extra centroids to achieve a better local slope estimation (Z-tilt instead of G-tilt) or reconstruct wavefront from multiple sources is a simple sorting procedure is beyond the scope of this paper. Here, we try to render a traditional array of centroids corresponding to the lenslet array from the SCoG detected centroids through the Segmentation module using the following steps.

3.2.1. Centroid Sorting Module

It is assumed that all the detected centroids in one sub-aperture truly belong to this sub-aperture. In other words, there are no spots crossing lenslet boundaries due to strong aberration. In order to associate a detected centroid to its sub-aperture, Equation (7) was used to calculate the row and column index of the sub-aperture. When the number of pixels per lenslet window and integer of power is two, the division can be simplified to bit shifts.

3.2.2. Treatment of Multiple Centroids and Missing Centroid

As discussed in Section 2.3, there are different ways to treat the scenarios where multiple centroids or no centroid is detected in some sub-apertures. For the current FPGA implementation, the average of all the centroids within a sub-aperture is used to represent an overall displacement for the multiple centroids situation. If a centroid is missing from a sub-aperture, then the centroid position from previous frame is inherited, which can trace back to an initial default reference centroid positions for each sub-aperture.

Once the centroids for the last sub-aperture in a row have been examined following by the slope calculation, the partial wavefront reconstruction can start immediately. The above segmentation steps can be represented by the pseudocode in Algorithm 1:

Algorithm 1: Pseudocode for *sorting–processing–reordering*.

```
Data: stream centroid C
Result: ordered centroid for each sub-aperture A
initialization;
foreach stream centroid C do
    r := floor(C_x/w);
    c := floor(C_y/w);
    append C to centroid list of A(r,c);
    if time of the last possible C in A then
        if no centroid in A(r,c) then
            Assign the centroid from previous frame;
        else if multiple centroids in A(r,c) then
            Average all the centroids;
        if last A in a MLA row then
            reorder all the centroids in this row;
        end
    end
end
```

3.3. Slope Calculation Module

After the SCoG and Segmentation modules, there will be an averaged centroid for each sub-aperture similar to conventional CoG-based methods. The SLOPE module calculates local slopes of wavefront in each sub-aperture according to Equation (1). An initial reference grid is stored in this module, where the reference center is designed to be in the center of each sub-aperture. This module is also capable of updating the reference centroids grid by averaging a certain frames of centroids data. This is particularly useful when applying the wavefront sensing system to an imaging system where a flat wavefront is not accessible. Therefore, the reference grid must be calculated by averaging the centroids from a number of long exposure images.

3.4. Wavefront Reconstruction Module

The wavefront reconstruction module essentially computes a matrix multiplication between the inverse matrix and the slope vectors according to Equation (12). As the slopes for each sub-aperture are measured, they are multiplied by their corresponding matrix elements and accumulated with previous multiplication. The calibration matrix E

is determined once the geometric configurations is fixed and therefore can be calculated in advance. When the slopes of all the sub-apertures in the same row are measured and multiplied, the next section of the matrix corresponding to the slopes of next row of sub-apertures are loaded. The loading of new matrix elements is controlled by a finite state machine (FSM).

In Figure 6, different areas related with the SHWFS are shown. The gray sensor area includes all the selected active imaging sensor pixels, on which the SCoG is operated. The resulted detected centroids from the SCoG module are represented on this sensor coordinate system. Since most AO systems have either a circular or annular Pupil, a square ROI was chosen to include the Pupil. After removing the ROI origin offset (os_x, os_y) from the centroids calculated by the SCoG module, the local centroids were obtained on the ROI coordinate system. Following that, the segmentation module can sort the centroids to their associated sub-apertures, followed by the slope module to obtain the average slope on individual sub-apertures. For the wavefront reconstruction, only slopes on those valid sub-apertures (green) were used. A binary sub-aperture mask indicating the validness of an sub-aperture was used to decide whether to accumulate matrix multiplication result to the partial wavefront reconstruction in the RECON module.

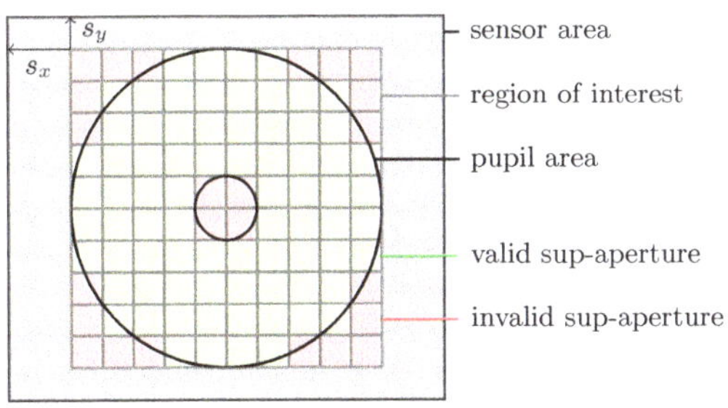

Figure 6. Illustration of various regions on the sensor area. The gray area is the sensor area on which the stream-based CoG operates. The region of interest is selected such that it covers the pupil of the imaging system. The green segments indicate those valid MLA lenslets within the pupil, while the red segments are invalid sub-apertures for wavefront reconstruction.

4. Results

4.1. FPGA Design

The hardware platform used to implement the proposed Shack–Hartmann wavefront sensor using the stream-based CoG method is shown in Figure 7. The FPGA board is a Trenz Electronic TE0712 core board (TE0712-01-200-2C), which includes a Xilinx 7 series FPGA Artix-7 chip XC7A200T.

Figure 7. Hardware platform used to implement the SCoG-based SHWFS.

The FPGA implementation was developed in the Xilinx Vivado Design Suite, with most parts of the different modules designed using VHDL directly. Some common IP blocks, such as FIFOs, Dividers, and Block Memories provided by Xilinx, have also been used in various places. This implementation strategy reduces the migration complexities and difficulties when re-targeting the design to a different FPGA platform.

The arithmetic operations starting from the calculation of the sub-pixel shift use fixed-point numbers. The absolute centroid position has a precision of 12 integer bits and 8 fractional bits, while the slope calculation uses 1 sign bit and 16 fractional bits. The intermediate calculations use full precision, i.e., integer and fractional bits are extended accordingly. The final Zernike amplitudes are rounded to 1 sign bit, 7 integer bits, and 8 fractional bits.

4.1.1. Resource Usage

The resource usage for the SCoG module, centroids segmentation module, slope calculation module, and modal wavefront reconstruction module for a SHWFS with 10×10 sub-apertures, each having 30×30 pixels. as is shown in Table 2. The filter size is 5×5 and the wavefront reconstruction has been performed on the first 9 Zernike modes based on Noll's notation. The SEG module uses more LUT RAMs than other components because centroids from previous frame are buffered to solve the possible missing centroid situations in some sub-apertures. The RECON module uses the most Block RAMs only because the calibration matrix are stored inside this module instead of external RAM.

Table 2. Resource usage of different modules on Artix 7 XC7A200T.

Resources	SCoG	SEG	Slope	RECON	Total
Slice LUT	1554	7744	33	3731	13,062 (10%)
Slice REG	3567	16,184	19	10,100	29,870 (11%)
Block RAM	3.5	0	0	26	29.5 (8.1%)
DSPs	0	0	2	18	20 (2.7%)

The resource usage for the SCoG module with four different filter sizes is shown in Figure 8. The implementation for slope calculation and wavefront reconstruction are common and similar to those in the literature [27,28] and therefore are not shown here separately. As the filter size increases, the slice usages as LUT and registers increase linearly. The Block RAMs are mainly used to buffer a certain number of rows pixels depending on the filter size, and therefore also increase linearly.

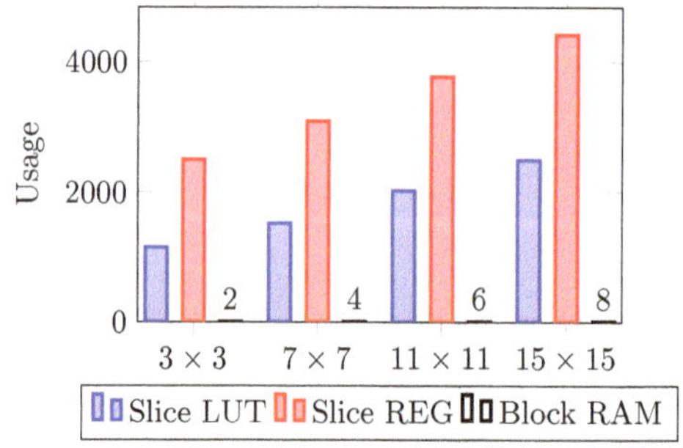

Figure 8. Resource usage for the SCoG module with four different filter sizes. Contributing resources for each filter size include Slice LUT, Block RAM, and Slice Registers.

4.1.2. Latency

The various time delays from the pixel read to the calculation of wavefront are shown in Figure 9. The SCoG computes the centroid on each incoming pixel and the time delay from a when pixel is read to its associated CoG value being calculated t_{SCoG} is:

$$t_{SCoG} = M \times T_{ROW} + M \times T_{CLK} + c \times T_{CLK} \tag{15}$$

where M is half of the filter size and c is a constant related to the delays caused by pipe-line implementation and division. The detected centroids will be sorted to a sub-aperture according to Algorithm 1, which introduces a delay t_{Sort} mostly due to divisions. When the CoG value of the last pixel in a row of sub-apertures is sorted, multiple centroids (if presented) in each individual sub-aperture will be averaged after a further delay of t_{SEG}. At this point, each sub-aperture will have its centroid estimation and the following timing delay is similar to those reported in [28]. The slope calculation can be started as soon as the segmentation is done, as the operation is basically a subtraction of the sub-aperture centroid with its reference center and following multiplication with a factor related with lenslet f-number. The time delay from the segmentation to the slope calculation is denoted as t_{SLOPE}.

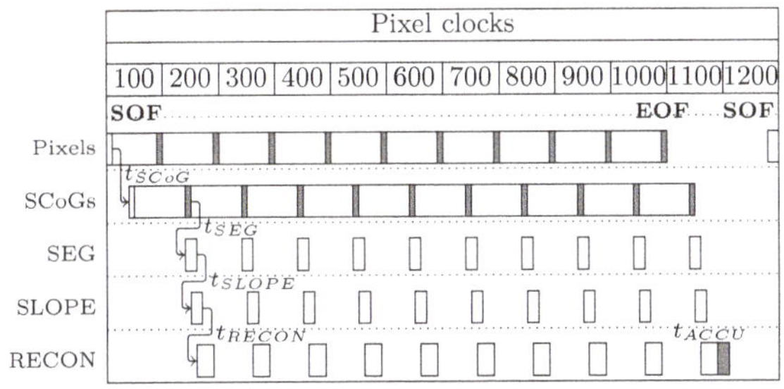

Figure 9. Time delay between different modules in the SHWFS implementation. Note that the whole wavefront sensing operation finishes shortly after End-Of-Frame (EOF) and before the next Start-Of-Frame (SOF).

The matrix multiplication for the wavefront reconstruction starts as soon as the slopes of all the sub-apertures in the same row becomes available and introduces a time delay of t_{RECON} until the partial reconstruction finishes. As the multiplication on the last row finishes, a time of t_{ACCU} is required to finish the final accumulation and serialization of the Zernike coefficients vectors.

With a larger filter size, the closure of the timing constraint becomes more challenging, even with fully pipe-lined implementation as described in Appendix A. In our implementation, the timing constraint for a clock running at 100 MHz can be met with a maximum filter size of 65 pixels. Considering the micro lenslet array has a pitch size ranging from 100 µm to 500 µm (equivalent f-number 20 to 50) and the common pixel size for CMOS imaging sensors of 5 to 10 µm, the spot size for a wavelength of 500 nm is about 4 to 20 pixels and each sub-aperture $\mathcal{A}$ is up to 50×50 pixels. Therefore, the timing constraint with this maximum filter size is sufficient for most of the SHWFS for astronomy AO systems. We note that most SHWFS limit the spot size to maximize the detectability, but meteorology look for better precision with the luxury of more light.

The appealing low latency and high accuracy of our proposed Shack–Hartmann WFS is only possible because of the deep exploitation of the parallelism power offered by FPGA devices. Due to the heavy computation overload introduced by the per-pixel convolution operations in Equation (3), CPU-based implementation will suffer from significant time delay. For a configuration of 10×10 sub-apertures with each sub-aperture consisting of 30×30 pixels and an implementation running at 50 MHz, the latency from the end of the frame to the finish of wavefront reconstruction to 9 Zernike terms is only 820 ns. The same software implementation of the Shack–Hartmann WFS, getting the benefits of CoG calculation at each pixel, written in Python 3.7 running on Ubuntu 18.04 (Intel Core i7-8750 CPU 2.20 GHz × 12, 32 GB RAM) (without exploiting specific optimisation like multithread programming) takes about 2.855 s (SCoG 2.85 s, SEG 0.19 ms, SLOPE, and RECON 4.77 ms). Image fetching time from memory in the CPU implementation has been omitted in the above comparison. The reader should therefore note that it is not sensible to implement the streaming per-pixel estimation of SCoG on a traditional CPU architecture for real-time applications. The latency of our FPGA-based Shack–Hartmann WFS implementation using SCoG for centroiding estimation is on par with similar FPGA implementation using TCoG [28] (740 ns) and is much smaller than the GPU-based centroiding extraction [22] (2 ms) and CPU-based Shack–Hartmann WFS implementation [32] (40 µs excluding transfer time).

4.2. Experimental Results

To evaluate the effectiveness of the hardware implementation of the proposed SHWFS using SCoG algorithm, artificial SHWFS spots images were generated from numerical simulation and stored inside the FPGA ROM. Following that, a camera module written in HDL was used to mimic an image sensor by reading the stored image and outputting common CMOS sensor signals such as pixel clock, frame/line valid, and pixel data bus. This arrangement allows the experiments to still run on the hardware platform and quantify the system performance in a reproducible manner.

4.2.1. Centroiding Performance

Comparison of the SCoG method with other CoG-based and cross-correlation methods for measuring the spot displacement under various noise scenarios has been reported in our previous research [33]. The focus here is to confirm that the argument of SCoG is superior to conventional CoG method still holds with the fixed-point number representation being used in various modules. The spot profile from the lenslet is modeled using the following 2D Gaussian function:

$$P(x,y) = \frac{N_{ph}}{2\pi\sigma_{spot}^2} \exp\left[-\frac{(x-x_0)^2+(y-y_0)^2}{2\sigma_{spot}^2}\right] \tag{16}$$

where (x_0, y_0) represents the ground truth of the spot center and N_{ph} is the number of photons. σ_{spot} is the standard deviation and the FWHM of the spot is $2\sqrt{2\ln(2)}\sigma_{spot}$.

A Gaussian spot with a FWHM of 2 pixels and 100 photons moving across a sub-aperture of 30 pixels is simulated as shown in Figure 10. On top of the Gaussian signal, photon (Poisson) noise and sensor readout noise are introduced as well. The readout noise is modelled by a Gaussian distribution with a mean value of 0.8 e^- and standard deviation of 0.2 e^- under assumption of 100% quantum efficiency. The final signal is then digitized to 8 bits. The centroids estimation results from a conventional TCoG with a thresholding value of 15 and the SCoG with a filter size of 5 pixels are shown in Figure 10. While the accuracy of the TCoG varies and at some sampling position results in a totally wrong estimation, the SCoG always has a close agreement with the ground truth along the path. The error variance of the centroid estimation from TCoG is about 1.795 Pixel^2, while the number from SCoG estimation is only about 0.002 Pixel^2.

To evaluate a case where traditional CoG methods fail, in Figure 11 we present a spots pattern from a random Kolmogorov phase screen with r_0 of 14 cm. A telescope with a diameter of 4.2 m, which leads to $D/r_0 = 30$, is used. The pupil was sampled into 10×10 sub-apertures by a MLA with pitch size of 42 cm. Here, r_0 is smaller than the MLA pitch size, and therefore scintillation is expected. All the detected centroids together with the centroid representation for each sub-aperture after segmentation are annotated on the spots in blue and red colors. Noticeably, the spot breaks as the local turbulence is too strong, i.e., r_0 is less than the lenslet size, as shown in the two inset sub-apertures, and the SCoG is able to identify these individual broken spots. As stated in the previous section, the average of these centroids was used to represent the mean local wavefront. However, they could be used in other, better ways [36] to yield a more precise wavefront reconstruction with further study. The SCoG works here, despite scintillation and the accompanying signal spread.

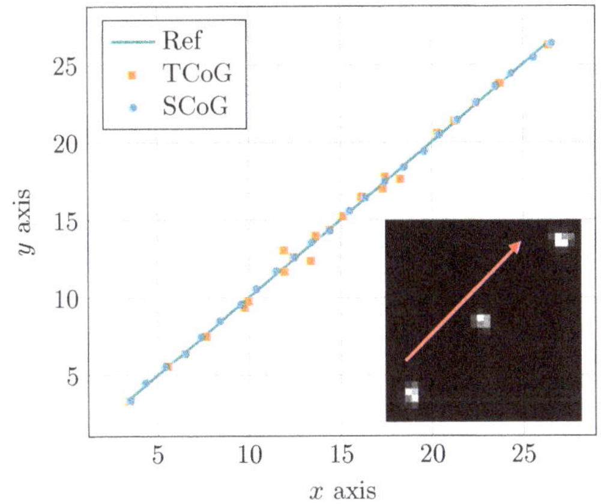

Figure 10. Comparison of hardware implemented SCoG with a traditional thresholding CoG (TCoG) for detecting a moving Gaussian spot.

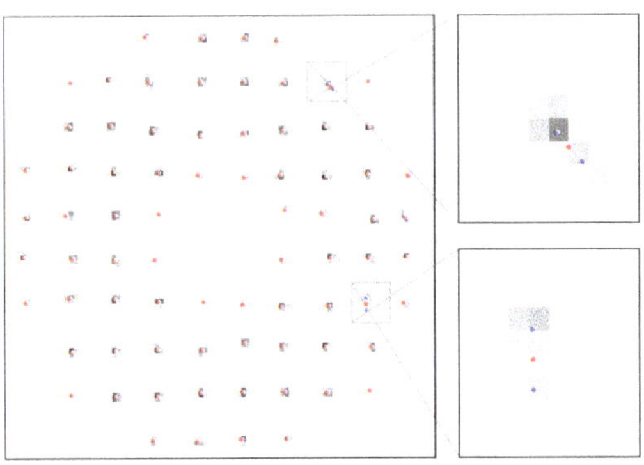

Figure 11. All the detected centroids (blue) and average ones for each sub-apertures (red) are annotated on the SHWFS spots pattern (invalid sub-apertures are not shown here). Color for the spots is inverted for better visibility.

4.2.2. Wavefront Reconstruction Results

In a similar way, to demonstrate the performance of the whole SHWFS design including the wavefront reconstruction, the SHWFS spots pattern was generated numerically from a known phase screen as shown in the top left in Figure 12. The wavefront is randomly generated by a combination of the first 9 Zernike modes, after setting a wavefront error budget of [0 nm, 100 nm, 100 nm, 36 nm, 36 nm, 36 nm, 6 nm, 6 nm, and 6 nm] at a wavelength of 635 nm and over a pupil with diameter of 1.5 mm. The MLA used to sample the wavefront has a pitch size of 150 µm and focal length of 4.1 mm, which models the Thorlabs MLA150-5C micro lenslet array. The pixel size of the detector is set to be 5 µm and therefore there are 10×10 sub-apertures and each sub-aperture has 30×30 pixels. The phase residual between the reconstructed wavefront and the original wavefront is shown in the top right in Figure 12 and the modal reconstruction has a RMSE of 0.023 rad. A comparison of the Zernike coefficients from the original and reconstructed wavefront is given in the bottom in Figure 12.

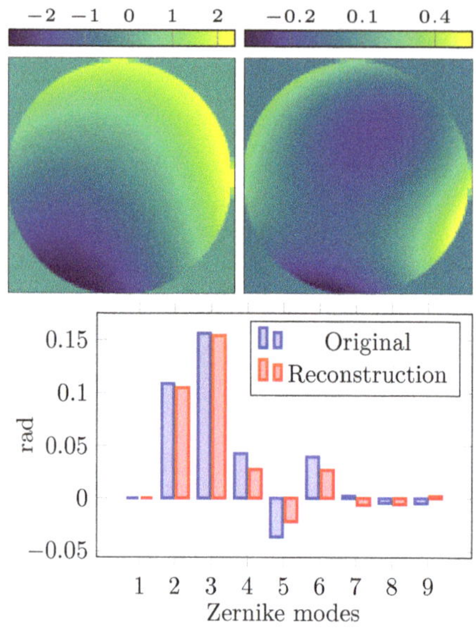

Figure 12. Top left: original phase; Top right: phase residual; Bottom: comparison of original and reconstructed Zernike coefficients using Noll's index notation [37].

5. Discussion

In this paper, we have presented a complete hardware implementation of Shack–Hartmann wavefront sensor in a modular design. The spots displacement was measured using the previously reported Stream-based CoG algorithm, which reduces the centroid estimation error when the signal is cut by the sub-aperture boundary and noise-induced error by using a spot-matched floating CoG window. The slope calculation and wavefront reconstruction using a least-square fit method starts immediately as long as the required sub-aperture measurements are ready. The result is a very low-latency and real-time wavefront sensing system with superior performance to conventional CoG-based SHWFS facilitated by the parallelism power from FPGA devices.

The FPGA implementation of Stream-based Center-of-Gravity algorithm can be modified without much effort to adapt other filter-like algorithms, for example, the cross-correlation for determining image shifts under extended sources proposed by Poyneer [6]. The First Fourier coefficient (1FC) algorithm [38] examines the phase symmetry in the Fourier domain to evaluate the spot shifts. Both the CCF and 1FC algorithms can be modified to their streamed version with minimal effort based on the implementation presented in this work.

The current treatment of multiple centroids identified in one sub-aperture is to average them, which can be further improved by other algorithm events to increase the dynamic range such as [36]. Future work also includes testing the proposed SHWFS implementation with the on-board image sensor in laboratory optics systems and eventually including the wavefront correction component, such as a deformable mirror, to realize a closed-loop AO system.

Author Contributions: Conceptualization, F.K. and A.L.; methodology, F.K., M.C.P. and A.L.; software, F.K.; validation, F.K., M.C.P. and A.L. formal analysis, F.K.; investigation, F.K., M.C.P. and A.L.; resources, A.L.; data curation, F.K.; writing—original draft preparation, F.K.; writing—review and editing, M.C.P. and A.L.; visualization, F.K.; supervision, M.C.P. and A.L.; project administration, A.L.; funding acquisition, A.L. All authors have read and agreed to the published version of the manuscript.

Funding: Parts of this research were funded by US AFOSR grant FA9550-18-1-0471.

Data Availability Statement: All data supporting this publication is held by UNSW in accordance with its Handling Research Material & Data Procedure.

Conflicts of Interest: The authors declare no conflict of interest.

Appendix A. Pipe-Lined Filter Implementation

The filters chosen for the SCoG algorithm are symmetric and separable, which is helpful to reduce the resource usage. For instance, $F_x(m,n)$ can be separated into two 1-D filters as following:

$$F_x(m,n) = \begin{bmatrix} 1 \\ 1 \\ \vdots \\ 1 \end{bmatrix} \begin{bmatrix} -M & -M+1 & \dots & M \end{bmatrix} \tag{A1}$$

Similarly, $F_y(m,n)$ can be separated by:

$$F_y(m,n) = \begin{bmatrix} -M \\ -M+1 \\ \vdots \\ M \end{bmatrix} \begin{bmatrix} 1 & 1 & \dots & 1 \end{bmatrix} \tag{A2}$$

The column vector is applied to each incoming column of pixels first, followed by a pipeline adder for the row vector. However, the multiple sum operations after the multiplication of the vertical vectors lead to complicated combination logic between two synchronous FFs and as a result, difficult timing closure. One solution to improve the timing performance of this path is to break down the operations to be completed in more clock cycles, i.e., a pipelined implementation as shown in Figure A1. Because the 1D filter $[-M - M+1 \dots M]^T$ in Equation (A2) is also vertically symmetrical, it is possible to further reduce the number of multipliers and delay FFs by summing the symmetrical pixels first. Similar optimization can be applied to the horizontal 1D filter operation. However, as our target FPGA device has plenty of resources for our current implementation, we have not done such optimization. The output after the column vector, DV, is then passed to be multiplied by the horizontal vector.

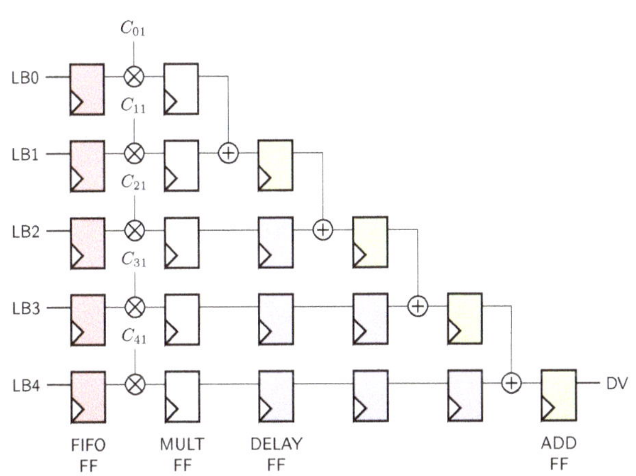

Figure A1. A fully pipe-lined design for the vertical filter operation. LB is the output of different line buffers and DV represents the output after applying the column vector.

The implementation of the filter with the horizontal vector is a bit different as the product of the previous vertical filter operation is present sequentially with each clock cycle. It is straightforward to think that we need to use registers to buffer the data, and then do the pipe-lined multiplication and addition as described for the vertical filtering. However, a better way, in terms of simplicity and resource usage, to implement this filter is to use the Transpose structure, which is a common technique in the design of finite impulse response (FIR) filter. As shown in Figure 4, the input DV, is multiplied by all the coefficients of the horizontal vector on each clock cycle. The results of multiplications are added and propagate through a chain of registers, with the result from the later coefficient at the furthest FF. The elegance of this Transpose structure, is that the adder chain can be easily placed and routed because it matches with the Xilinx FPGA structure, where arithmetic functions are placed in columns. In fact, the adder and register can be absorbed to the DSP48A slices which are used for the multiplication, which not only reduces the resource usage but also improves the speed.

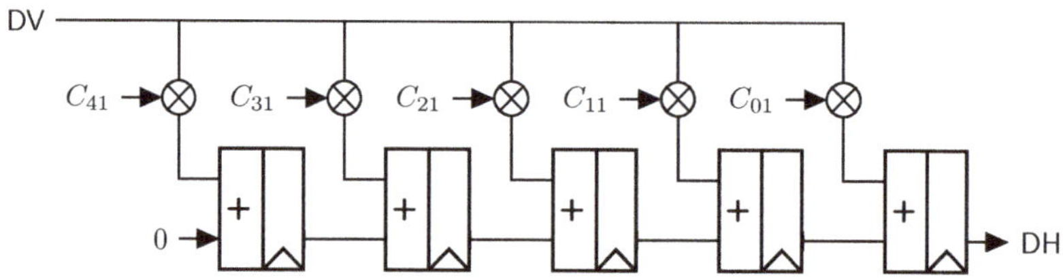

Figure A2. The horizontal filter implemented with a Transpose structure which takes the output of the column vector operation, DV, and produces DH in its output.

References

1. Platt, B.C.; Shack, R. History and Principles of Shack-Hartmann Wavefront Sensing. *J. Refract. Surg.* **2001**, *17*, S573–S577. [CrossRef] [PubMed]
2. Leroux, C.; Dainty, C. Estimation of Centroid Positions with a Matched-Filter Algorithm: Relevance for Aberrometry of the Eye. *Opt. Express* **2010**, *18*, 1197–1206. [CrossRef] [PubMed]
3. van Dam, M.A.; Lane, R.G. Wave-Front Slope Estimation. *J. Opt. Soc. Am. A* **2000**, *17*, 1319–1324. [CrossRef] [PubMed]
4. Barrett, H.H.; Dainty, C.; Lara, D. Maximum-Likelihood Methods in Wavefront Sensing: Stochastic Models and Likelihood Functions. *J. Opt. Soc. Am. A* **2007**, *24*, 391–414. [CrossRef]
5. Michau, V.; Rousset, G.; Fontanella, J. Wavefront Sensing from Extended Sources. In *Real Time and Post Facto Solar Image Correction;* Radick, R.R., Ed.; SAO/NASA Astrophysics Data System (ADS): online, 1993; p. 124.
6. Poyneer, L.A. Scene-Based Shack-Hartmann Wave-Front Sensing: Analysis and Simulation. *Appl. Opt.* **2003**, *42*, 5807–5815. [CrossRef]
7. Knutsson, P.A.; Owner-Petersen, M.; Dainty, C. Extended Object Wavefront Sensing Based on the Correlation Spectrum Phase. *Opt. Express* **2005**, *13*, 9527–9536. [CrossRef]
8. Thomas, S.; Fusco, T.; Tokovinin, A.; Nicolle, M.; Michau, V.; Rousset, G. Comparison of Centroid Computation Algorithms in a Shack–Hartmann Sensor. *Mon. Not. R. Astron. Soc.* **2006**, *371*, 323–336. [CrossRef]
9. Rousset, G. Wave-front sensors. In *Adaptive Optics in Astronomy;* Roddier, F., Ed.; Cambridge University Press: Cambridge, UK, 2004; Chapter 5, pp. 91–130.
10. Arines, J.; Ares, J. Minimum Variance Centroid Thresholding. *Opt. Lett.* **2002**, *27*, 497–499. [CrossRef]
11. Li, X.; Li, X.; Wang, C. Optimum Threshold Selection Method of Centroid Computation for Gaussian Spot. *Proc. SPIE* **2015**, *9675*, 967517. [CrossRef]
12. Nicolle, M.; Fusco, T.; Rousset, G.; Michau, V. Improvement of Shack-Hartmann Wave-Front Sensor Measurement for Extreme Adaptive Optics. *Opt. Lett.* **2004**, *29*, 2743–2745. [CrossRef]
13. Baker, K.L.; Moallem, M.M. Iteratively Weighted Centroiding for Shack-Hartmann Wave-Front Sensors. *Opt. Express* **2007**, *15*, 5147–5159. [CrossRef]
14. Ma, X.; Rao, C.; Zheng, H. Error Analysis of CCD-based Point Source Centroid Computation Under the Background Light. *Opt. Express* **2009**, *17*, 8525–8541. [CrossRef]
15. Yin, X.; Li, X.; Zhao, L.; Fang, Z. Automatic centroid detection for Shack-Hartmann Wavefront sensor. In Proceedings of the 2009 IEEE/ASME International Conference on Advanced Intelligent Mechatronics, Singapore, 14–17 July 2009; pp. 1986–1991. [CrossRef]
16. Spiricon. *Hartmann Wavefront Analyzer Tutorial;* Spiricon, Inc.: North Logan, UT, USA, 2004.

17. Bezzubik, V.; Belashenkov, N.; Soloviev, O.; Vasilyev, V.; Vdovin, G. Hartmann-Shack Wavefront Reconstruction with Bitmap Image Processing. *Optics Lett.* **2020**, *45*, 972. [CrossRef]
18. Li, Z.; Li, X. Centroid Computation for Shack-Hartmann Wavefront Sensor in Extreme Situations Based on Artificial Neural Networks. *Opt. Express* **2018**, *26*, 31675–31692. [CrossRef]
19. Hu, L.; Hu, S.; Gong, W.; Si, K. Learning-Based Shack-Hartmann Wavefront Sensor for High-Order Aberration Detection. *Opt. Express* **2019**, *27*, 33504. [CrossRef]
20. Lechner, D.; Zepp, A.; Eichhorn, M.; Gładysz, S. Adaptable Shack-Hartmann Wavefront Sensor with Diffractive Lenslet Arrays to Mitigate the Effects of Scintillation. *Opt. Express* **2020**, *28*, 36188. [CrossRef]
21. Talmi, A.; Ribak, E.N. Direct Demodulation of Hartmann-Shack Patterns. *J. Opt. Soc. Am. A* **2004**, *21*, 632–639. [CrossRef]
22. Mocci, J.; Busato, F.; Bombieri, N.; Bonora, S.; Muradore, R. Efficient Implementation of the Shack-Hartmann Centroid Extraction for Edge Computing. *J. Opt. Soc. Am. A* **2020**, *37*, 1548. [CrossRef]
23. Mompeán, J.; Aragón, J.L.; Prieto, P.M.; Artal, P. GPU-based Processing of Hartmann-Shack Images for Accurate and High-Speed Ocular Wavefront Sensing. *Future Gener. Comput. Syst.* **2019**, *91*, 177–190. [CrossRef]
24. Goodsell, S.J.; Fedrigo, E.; Dipper, N.A.; Donaldson, R.; Geng, D.; Myers, R.M.; Saunter, C.D.; Soenke, C. FPGA developments for the SPARTA project. *Proc. SPIE* **2005**, *5903*, 59030G. [CrossRef]
25. Mauch, S.; Reger, J.; Reinlein, C.; Appelfelder, M.; Goy, M.; Beckert, E.; Tünnermann, A. FPGA-accelerated adaptive optics wavefront control. *Proc. SPIE* **2014**, *8978*, 897802. [CrossRef]
26. Mauch, S.; Barth, A.; Reger, J.; Reinlein, C.; Appelfelder, M.; Beckert, E. FPGA-accelerated adaptive optics wavefront control part II. *Proc. SPIE* **2015**, *9343*, 93430Y. [CrossRef]
27. Mauch, S.; Reger, J. Real-Time Spot Detection and Ordering for a Shack-Hartmann Wavefront Sensor with a Low-Cost FPGA. *IEEE Trans. Instrum. Meas.* **2014**, *63*, 2379–2386. [CrossRef]
28. Thier, M.; Paris, R.; Thurner, T.; Schitter, G. Low-Latency Shack-Hartmann Wavefront Sensor Based on an Industrial Smart Camera. *IEEE Trans. Instrum. Meas.* **2013**, *62*, 1241–1249. [CrossRef]
29. Kincses, Z.; Orzó, L.; Nagy, Z.; Mező, G.; Szolgay, P. High-Speed, SAD Based Wavefront Sensor Architecture Implementation on FPGA. *J. Signal Process. Syst.* **2011**, *64*, 279–290. [CrossRef]
30. Saunter, C.D.; Love, G.D.; Johns, M.; Holmes, J. FPGA Technology for High-Speed Low-Cost Adaptive Optics. *Proc. SPIE* **2005**, *6018*, 60181G. [CrossRef]
31. Cegarra Polo, M. Adaptive Optics For Small Aperture Telescopes. Ph.D. Thesis, The University of New South Wales, Sydney, NSW, Australia, 2015.
32. Mocci, J.; Quintavalla, M.; Trestino, C.; Bonora, S.; Muradore, R. A Multiplatform Cpu-Based Architecture for Cost-Effective Adaptive Optics Systems. *IEEE Trans. Ind. Inform.* **2018**, *14*, 4431–4439. [CrossRef]
33. Kong, F.; Polo, M.C.; Lambert, A. Centroid Estimation for a Shack-Hartmann Wavefront Sensor Based on Stream Processing. *Appl. Opt.* **2017**, *56*, 6466–6475. [CrossRef]
34. Hardy, J. *Adaptive Optics for Astronomical Telescopes*; Oxford Series in Optical and Imaging Sciences; Oxford University Press: Oxford, UK, 1998.
35. Poyneer, L.A.; Gavel, D.T.; Brase, J.M. Fast Wave-Front Reconstruction in Large Adaptive Optics Systems with Use of the Fourier Transform. *J. Opt. Soc. Am. A* **2002**, *19*, 2100. [CrossRef]
36. Leroux, C.; Dainty, C. A Simple and Robust Method To Extend the Dynamic Range of an Aberrometer. *Opt. Express* **2009**, *17*, 19055–19061. [CrossRef]
37. Noll, R.J. Zernike Polynomials and Atmospheric Turbulence. *J. Opt. Soc. Am.* **1976**, *66*, 207–211. [CrossRef]
38. Lambert, A.; Cegarra Polo, M. Real-time algorithms implemented in hardware for centroiding in a Shack-Hartmann Sensor. In Proceedings of the Imaging and Applied Optics, Arlington, VI, USA, 7–11 June 2015; Optical Society of America: Washington, DC, USA, 2015; p. AOW3F.2. [CrossRef]

Disclaimer/Publisher's Note: The statements, opinions and data contained in all publications are solely those of the individual author(s) and contributor(s) and not of MDPI and/or the editor(s). MDPI and/or the editor(s) disclaim responsibility for any injury to people or property resulting from any ideas, methods, instructions or products referred to in the content.

electronics

Article

An Instruction-Driven Batch-Based High-Performance Resource-Efficient LSTM Accelerator on FPGA

Ning Mao [1,2], Haigang Yang [3,4,*] and Zhihong Huang [1,2]

1 Aerospace Information Research Institute, Chinese Academy of Sciences, Beijing 100190, China; maoning115@mails.ucas.ac.cn (N.M.); huangzhihong@mail.ie.ac.cn (Z.H.)
2 School of Electronic, Electrical and Communication Engineering, University of Chinese Academy of Sciences, Beijing 100094, China
3 School of Integrated Circuits, University of Chinese Academy of Sciences, Beijing 100049, China
4 Shandong Industrial Institute of Integrated Circuits Technology Ltd., Jinan 250001, China
* Correspondence: yanghg@mail.ie.ac.cn

Abstract: In recent years, long short-term memory (LSTM) has been used in many speech recognition tasks, due to its excellent performance. Due to a large amount of calculation and complex data dependencies of LSTM, it is often not so efficient to deploy on the field-programmable gate array (FPGA) platform. This paper proposes an LSTM accelerator, driven by a specific instruction set. The accelerator consists of a matrix multiplication unit and a post-processing unit. The matrix multiplication unit uses staggered timing of read data to reduce register usage. The post-processing unit can complete various calculations with only a small amount of digital signal processing (DSP) slices, through resource sharing, and at the same time, the memory footprint is reduced, through the well-designed data flow design. The accelerator is batch-based and capable of computing data from multiple users simultaneously. Since the calculation process of LSTM is divided into a sequence of instructions, it is feasible to execute multi-layer LSTM networks as well as large-scale LSTM networks. Experimental results show that our accelerator can achieve a performance of 2036 GOPS at 16-bit data precision, while having higher hardware utilization compared to previous work.

Keywords: LSTM; FPGA; resource efficient; accelerator

Citation: Mao, N.; Yang, H.; Huang, Z. An Instruction-Driven Batch-Based High-Performance Resource-Efficient LSTM Accelerator on FPGA. *Electronics* **2023**, *12*, 1731. https://doi.org/10.3390/electronics12071731

Academic Editors: Andres Upegui, Andrea Guerrieri and Laurent Gantel

Received: 27 February 2023
Revised: 4 April 2023
Accepted: 4 April 2023
Published: 5 April 2023

Copyright: © 2023 by the authors. Licensee MDPI, Basel, Switzerland. This article is an open access article distributed under the terms and conditions of the Creative Commons Attribution (CC BY) license (https://creativecommons.org/licenses/by/4.0/).

1. Introduction

In recent years, recurrent neural nets have been widely used in tasks such as speech recognition [1], due to their excellent performance. Long short-term memory (LSTM) is one of the most popular recurrent neural networks. A central processing unit (CPU) and graphics processing unit (GPU) are common LSTM hardware computing platforms. Due to a large number of calculations and complex data dependencies in LSTM, it is often not so efficient to calculate LSTM through CPU or GPU. When performing specific LSTM calculations, the utilization of CPU and GPU is usually relatively low. Due to the aforementioned drawbacks of CPU and GPU, some energy-efficient platforms are used as accelerators for LSTM forward inference, such as field-programmable gate arrays (FPGAs) and application-specific integrated circuits (ASICs). ASICs are highly energy efficient. For example, Google's TPUv4i [2] has a performance of up to 138 tera floating point operations per second (TFLOPS). However, ASICs are inflexible and expensive to manufacture. Specific ASIC chips may not keep up with the development of neural network algorithms.

FPGAs have achieved a good balance in terms of reconfigurability, flexibility, performance, and power consumption. At present, many researchers use FPGAs to accelerate LSTM [3–23]. Some works reduce the storage space of LSTM by compressing and quantizing the weight of LSTM, and then storing all the weight on the chip [9,11]. In work [9], the LSTM model is compressed using the block-circulant matrix technology so that the model parameters can be stored on the on-chip block random access memory (BRAM) of the FPGA. In [11], the bank-balanced sparsity method was used to reduce the number of parameters, so that all weights were stored on-chip in the small model. Some works store weights in off-chip memory and reduce bandwidth requirements through data reuse [16,19]. The authors of [19] split the weight matrix into multiple blocks, and each block could be used for a batch of input data, which increased data reuse. Addressing edge computing scenarios, the authors of [12] focused on using embedded FPGAs to accelerate lightweight LSTM. However, most works use FPGAs to accelerate an LSTM model, which cannot effectively accelerate some multi-layer LSTM. Most of them use different computing units to calculate different matrix multiplications of an LSTM, to those used to calculate operations such as element-wise multiplication and element-wise addition in LSTM.

In this paper, we propose an instruction-driven accelerator. For large LSTM and multi-layer LSTM, LSTM can be decomposed into multiple groups of small calculations through a series of instructions. Our hardware consists of matrix multiplication units and post-processing units, that compute element-wise multiplication, element-wise addition, etc. Several optimization techniques are used to improve performance and utilize resources efficiently.

The contributions of this work are summarized as follows.

- The matrix multiplication units in the accelerator are cascaded, to simultaneously compute multiple user input data. The matrix multiplication units reduce register usage through a time-staggered data readout strategy.
- In the post-processing unit, only two DSPs with resource sharing are used, to complete various types of calculations such as element-wise multiplication and batch normalization, which reduces the resource usage of the post-processing unit.
- A domain-specific instruction set is designed to compute complex operations in LSTM. A complex LSTM is executed by splitting it into a sequence of instructions.
- For a case study, experiments have been performed on the Xilinx Alevo U50 card. The results show that our design achieves a performance of 2036 giga operations per second (GOPS) and the utilization of the hardware reaches 86.1%.

The rest of the paper is organized as follows. Section 2 introduces the background. Section 3 presents the hardware architecture design. Section 4 describes the detailed instruction design. Section 5 gives an analysis of the experimental results. Section 6 concludes the paper.

2. Background

LSTM was first proposed in 1997 [24], and there have been many variants of LSTM since then. Google LSTM [25] is one variant that has been widely used. Therefore, without loss of generality, this paper uses Google LSTM as an example. Figure 1 shows the network structure of LSTM.

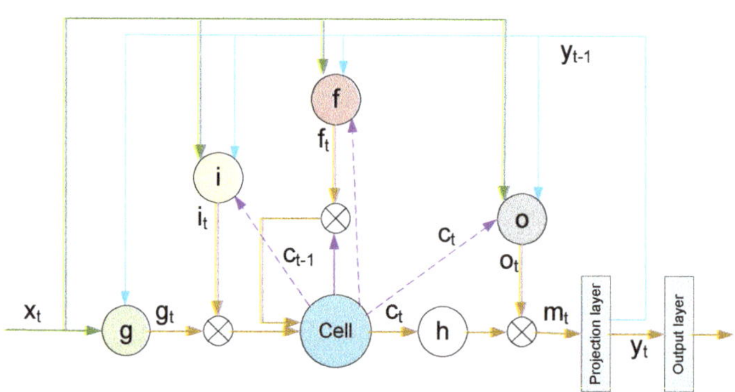

Figure 1. LSTM network structure.

Compared with the standard LSTM, Google LSTM has additional peephole connections and a projection layer. This structure has a better effect on deep networks. The input of an LSTM is a sequence $X = \{x_1; x_2; \ldots; x_t\}$; x_t represents the input vector at time t. The output of an LSTM is a sequence $Y = \{y_1; y_2; \ldots; y_t\}$; y_t denotes the output vector at time t. The operation of an LSTM can be expressed as:

$$g_t = \sigma(W_{gx}x_t + W_{gr}y_{t-1} + b_g). \tag{1}$$

$$i_t = \sigma(W_{ix}x_t + W_{ir}y_{t-1} + W_{ic} \odot c_{t-1} + b_i). \tag{2}$$

$$f_t = \sigma(W_{fx}x_t + W_{fr}y_{t-1} + W_{fc} \odot c_{t-1} + b_f). \tag{3}$$

$$c_t = f_t \odot c_{t-1} + g_t \odot i_t. \tag{4}$$

$$o_t = \sigma(W_{ox}x_t + W_{or}y_{t-1} + W_{oc} \odot c_t + b_o). \tag{5}$$

$$m_t = o_t \odot tanh(c_t). \tag{6}$$

$$y_t = W_{ym}m_t. \tag{7}$$

where $\boldsymbol{i}$, $\boldsymbol{f}$, $\boldsymbol{o}$, $\boldsymbol{c}$, $\boldsymbol{m}$ represent input gate, forget gate, output gate, cell state, and cell output, respectively. $\odot$ represents element-wise multiplication. $+$ denotes element-wise addition. $\boldsymbol{W}$ denotes represents the weight matrix. $\boldsymbol{b}$ denotes the bias used in the matrix multiplication operation. σ denotes the sigmoid activation function. $\boldsymbol{tanh}$ represents the tanh activation function.

The input gate controls the proportion of input information transmitted to the cell state at the current time step. The output gate controls the proportion of the cell state transmitted to the output. The forget gate controls the proportion of information forgotten and retained by the cell state. Cell state saves previous information. Because the number of units in the projection layer is less than that of the hidden units, the projection layer can control the total number of parameters, while allowing the number of hidden units to be increased. The output layer calculates the final output.

LSTM will save previous information and be able to learn the relationship between data at different times, so that LSTM can process sequential data such as voice data. Therefore, LSTM is widely used in tasks such as speech recognition, machine translation, etc.

Next, we will introduce some principles of hardware acceleration for computing. Common methods of hardware acceleration include pipelining, loop unrolling, loop tiling, and loop interchange, among others. Pipelining can increase the operating frequency of the system. Loop unrolling can improve parallelism during acceleration. In the case of insufficient on-chip storage resources, loop tiling can process part of the data at a time. Reasonable loop interchange can improve data reuse and optimize data movements and memory access. In computing tasks, multiple computing units of FPGA can be used for parallel processing. In the calculation process, improving the effective utilization of computing resources is helpful to the final performance.

3. Hardware Architecture

Our complete design consists of a hardware accelerator and corresponding instruction set design. Section 3 introduces the design of each module in the hardware part.

3.1. Overall Architecture

Our design consists of a host and a kernel. The host program is written in C++ and runs on the CPU, and the kernel program is written in Verilog and runs on the FPGA. As shown in Figure 2, data pre-processing, including feature extraction, and post-processing, including language models, are all calculated in the host, while LSTM computing is in the kernel. When the system starts to work, the host pre-processes the data first and then sends instructions, weights, parameters, the activation function table, and the pre-processed input data to high bandwidth memory (HBM). Then, the kernel starts executing the instructions. When the kernel finishes computing, the kernel notifies the host and puts the result on the HBM. Then, the host will perform the remaining calculations, such as the calculation of the language model, and notify the kernel to proceed with the next operation.

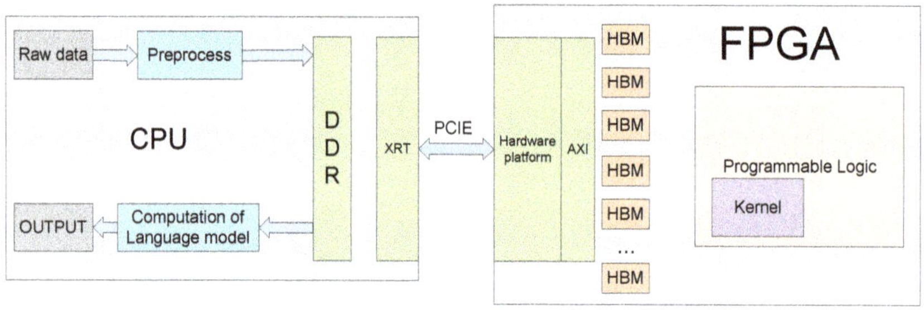

Figure 2. The overall architecture.

3.2. Kernel Architecture

The LSTM is calculated in the kernel, and the kernel in the FPGA completes most of the calculations of the entire system. The following describes the architecture of the kernel, as shown in Figure 3.

Instructions, weights, post-processing parameters, activation function tables, and input and output data are stored on HBM. Instructions, post-processing parameters and sigmoid tables each take up one HBM. The weight data takes up eight HBMs. The input and output data share an HBM. Inside the computing kernel, each HBM has a corresponding FIFO (first in first out). Each block of HBM uses an AXI interface for data reading and writing. When the kernel starts working, the instruction data are transferred from the HBM to the corresponding FIFO_S1. The state of the command FIFO_S1 is monitored. When the command FIFO_S1 is not empty, the data in the FIFO_S1 will be read and sent to FIFO_S2, and a signal to read data in other HBM will be pulled high. Next, the AXI1 interface to the AXI12 interface will read the data in the HBM, according to the information in the instruction, and transfer it to the corresponding FIFO. At the same time, the data in FIFO12 will be sent to ultra random access memory (URAM), which stores input data and intermediate results.

The calculation of the kernel does not need to wait for all the input data to enter the URAM, and the calculation of the kernel starts when sufficient input data is sent to the URAM. At this time, the instruction enters the kernel from the FIFO_S2, and the matrix multiplication unit will obtain data from the weight FIFO and URAM, according to the information in the instruction for calculation. When the matrix multiplication is finished, the result will be written into the BRAM, and then the post-processing unit will start the calculation. The post-processing unit will obtain the data from FIFO or BRAM, according to the information in the instruction, and obtain the parameters from the FIFO9, storing the parameters for the calculation. Because the two modules write BRAM at different times, the two modules perform calculations simultaneously. After the matrix multiplication unit completes the first set of data calculations, the post-processing unit performs the first set of data calculations and the second set of data enters the matrix multiplication unit.

As shown in Figure 3, the results calculated by the post-processing unit will be stored in different storage units. Which memory unit is written to, is determined by the information in the instruction. Finally, the data that needs to be written into HBM will be sent to FIFO first and then written into HBM. The proposed batch-based accelerator can process input data from multiple users simultaneously. The input data of multiple users is calculated simultaneously in the kernel, and a total of 32 groups of computing units perform calculations simultaneously. Since the 32 groups of computing units share the same weights and post-processing parameters, the weights and post-processing parameters are transferred between them through cascading. This calculation mode reduces the demand for external HBM storage bandwidth, by reusing weights.

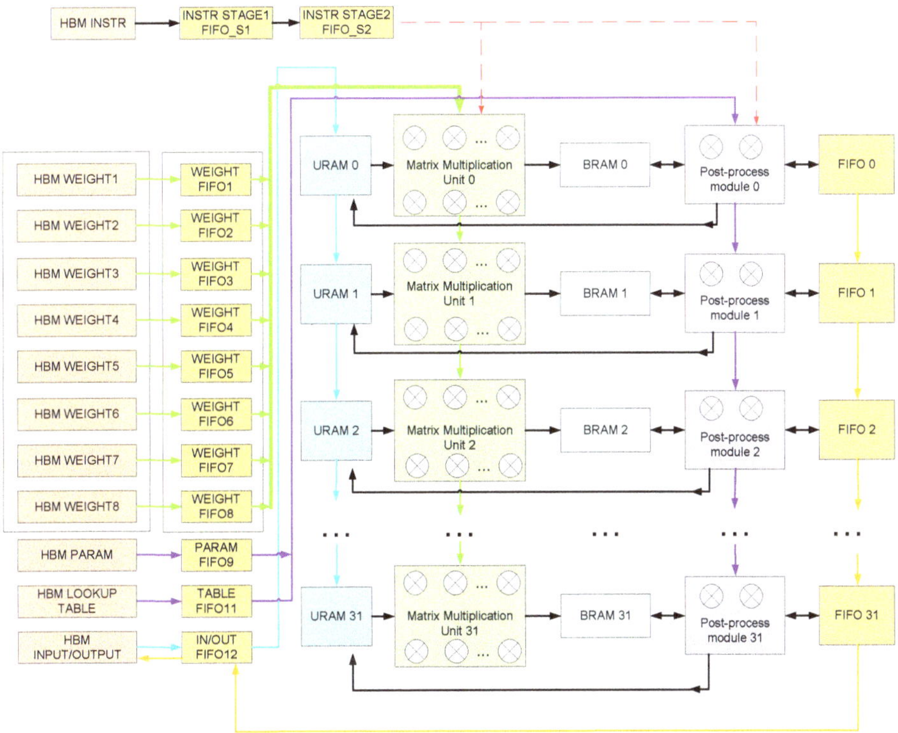

Figure 3. The architecture of the kernel (LSTM accelerator running on PL).

3.3. *Design of Matrix Multiplication Unit*

Most of the calculations in LSTM are matrix multiplications, and matrix multiplications have the greatest demand for computing resources. We design a matrix multiplication array composed of DSP to perform matrix multiplication. The detailed design of the matrix multiplication unit is shown in Figure 4 below.

In Figure 3, we take a 3 × 3 array as an example, and the actual array size is 8 × 8. There are 64 DSPs in total, 8 DSPs in each column as a group. Each group of DSPs calculates the same value in the output matrix. In this calculation mode, the eight groups of DSPs use the same input vector, so the eight groups of DSPs can share the same input vector, through cascading among different groups. This calculation method reduces the bandwidth requirement of the input vector to one-eighth of the calculation mode, without input data sharing. In a matrix multiplication unit, different DSPs use different weights, without using weight sharing. Because each group of eight DSPs computes the same output value, the eight values computed by the eight DSPs are added, to form a partial sum. A complete matrix multiplication operation will be performed multiple times by eight DSPs, and the partial sums obtained from the multiple operations will be accumulated. When a new set of data needs to be calculated, the data input port of the accumulator will be set to zero to perform a new calculation, as shown in Figure 4.

Since different DSPs in a group start calculations at staggered times, the value of the input vector will be registered through registers, and different DSPs in the same group will use different numbers of registers, as shown in Figure 4. For the reading of weight data, in a common design, the weight is read from the FIFO and then output through a series of registers, as shown in Figure 5. Since our design has many matrix multiplication units, the weight readout circuit will consume a lot of registers. This may cause difficulties in placement and routing, and result in a relatively low frequency for the final design. In order to reduce the use of registers, we have designed a specific data read timing, as shown in Figure 5. The read signal arrives at eight different weights, FIFO is staggered. In this mode, each DSP reduces the corresponding 4.5 registers on average. If each data is 16 bits and a matrix multiplication unit has 64 DSPs, then 64 × 16 × 4.5 = 4608 registers can be saved.

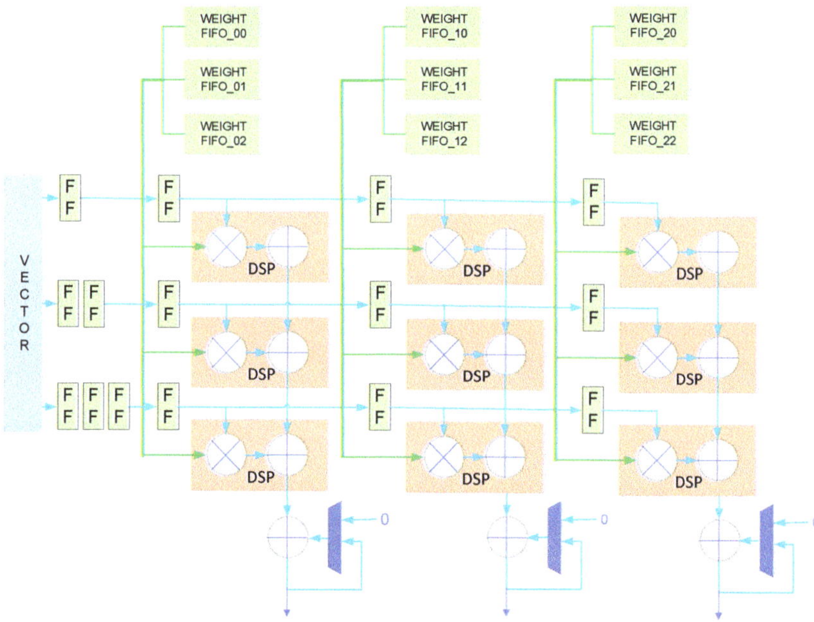

Figure 4. Design of matrix multiplication unit.

3.4. Design of Post-Processing Unit

After the matrix multiplication is completed, the result will be stored in the BRAM, and the post-processing unit will then obtain the data from the BRAM for calculation. The post-processing unit handles all operations except matrix multiplication, including element-wise multiplication, element-wise addition, batch normalization, etc. Although there is no batch normalization operation in LSTM, supporting batch normalization can broaden the applicability of the architecture, so that the accelerator can perform more types of calculations. In particular, in order to simplify the design of the matrix multiplication unit, the addition of bias operation in Equations (1)–(5) is also performed in the post-processing unit. In conventional operations, these operations require separate computing resources. Each type of calculation uses a dedicated computing resource. Since these operations are not performed at the same time, this will lead to a waste of computing resources. In order to improve resource utilization, we propose an architecture that utilizes two DSPs to perform all of the above operations. This is achieved through dynamic reconfiguration of the DSP. We make the DSP calculate different operations at different times by changing the operation code when the DSP is running. The detailed structure diagram is shown in Figure 6.

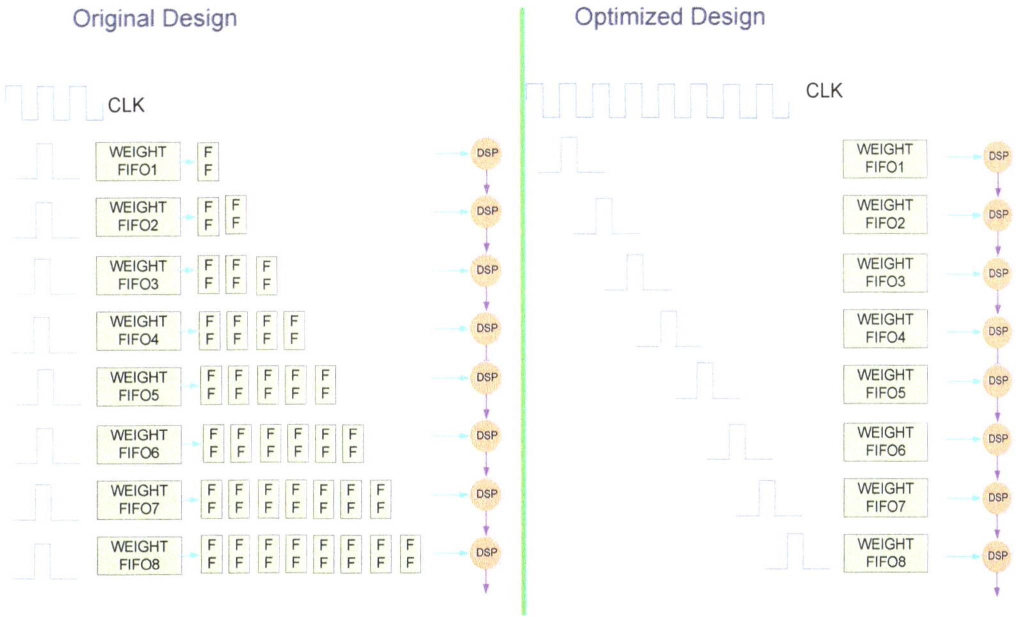

Figure 5. Optimized data read timing.

As shown in Figure 6, the post-processing parameters store the calculation type command, some weights, such as W_{ic}, W_{oc}, and the command that determines where the output is stored. Two DSPs are connected by cascading. Firstly, the type of operation performed by the two DSPs is determined by the command word, stored in the post-processing parameters. Different types of operations determine different operation codes. The input data of the two DSPs will be obtained from the BRAM, FIFO, and post-processing parameters through the multiplexer, and which data is to be selected is determined by the operation code. The shift module and the clamp module perform shift operations and truncate values, respectively.

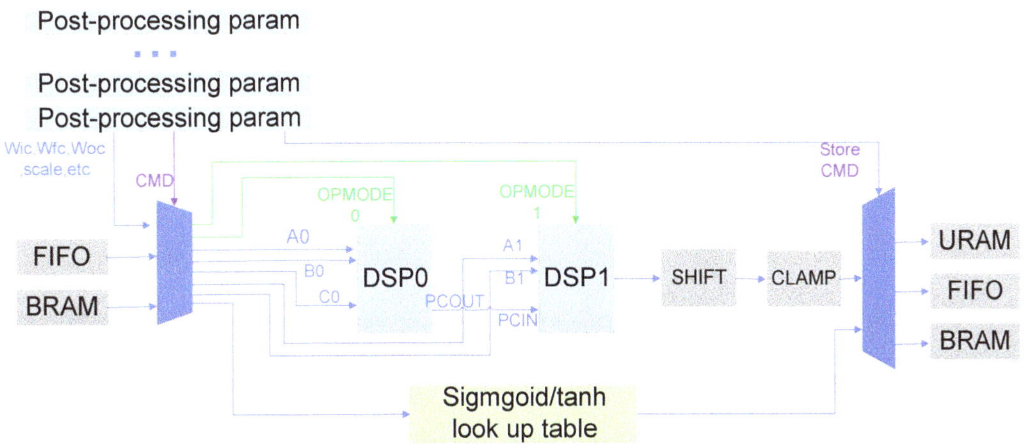

Figure 6. Structure of the post-processing unit.

Table 1 illustrates the configuration modes and opcodes of the DSP in different types of operations. Taking the addition of bias in matrix multiplication as an example, in the first stage, we will split the 36 bit data into two input ports A0 and B0 of DSP0, because the output result of matrix multiplication is 36 bit. The bias is placed on the C0 port and the operation completed by DSP0 is $\boldsymbol{result1}[35:0] = (S+C)((A0:B0)+C0)$. A colon indicates a bitwise data splicing operation. $S+C$ represent shift and clamp, respectively. In the second stage, we put $\boldsymbol{result1}[16:0]$ and $\boldsymbol{result}[35:17]$ on port A0 of DSP0 and port A1 of DSP1, respectively. The quantization parameter scale, representing the ratio between the fixed-point number and the actual floating-point number, is placed on the B ports of the two DSPs. At this time, the operation performed by the two DSPs is $\boldsymbol{result2}[15:0] = (S+C)(\boldsymbol{data}[16:0] \times \boldsymbol{scale} + \boldsymbol{data}[35:17] \times \boldsymbol{scale} << 17) = (S+C)(\boldsymbol{data}[35:0] \times \boldsymbol{scale})$.

Taking batch normalization as an example, data, gamma, and beta are, respectively, placed on ports A0, B0, and C0 of DSP0. In this mode, the operation completed by DSP is $\boldsymbol{result} = (S+C)(\boldsymbol{data1} * \boldsymbol{gamma} + \boldsymbol{beta})$. For element-wise multiplication and element-wise addition, the calculation process is similar. There are two types of element-wise multiplication in Table 1, which correspond to the two types in the LSTM formula. One type is data multiplied by weights, and the other is data multiplied by data. If these operations are calculated separately, using different DSPs, eight DSPs are required, whereas the proposed architecture reduces the number of DSPs required for these operations to two.

Table 1. Configuration modes and opcodes of the DSP.

Operations	Opmode0	Opmode1	A0	B0	C0	A1	B1
Element-wise multiplication type1	110000101	110010101	data1	weight	0	0	0
Element-wise multiplication type2	110000101	110010101	data1	data2	0	0	0
Element-wise addition	110000101	110010101	data1	scale1	data2	scale2	0
Batch normalization	110000101	110010101	data1	gamma	beta	0	0
Bias adding stage1	110000011	110010011	data[35:18]	data[17:0]	bias	0	0
Bias adding stage2	110000101	111010101	data[16:0]	scale	0	data[35:17]	scale

3.5. Design of Sigmoid/Tanh Module

The sigmoid and tanh functions are implemented in our architecture via a lookup table. The values in the lookup table are precomputed in software and loaded into the memory storing the lookup table in advance, before the activation function operation. Using the internal symmetry of the sigmoid function and the tanh function, we only need to store half of the data.

In order to improve the calculation efficiency, and avoid the activation function becoming a bottleneck, we designed a double buffer. Figure 7 shows a structural diagram of a double buffer that implements an activation function. A double buffer corresponds to a ping-pong operation. One buffer consists of two URAM and the double buffer consists of four URAM. The double buffer stores different lookup tables. For example, one buffer stores the sigmoid table while another buffer stores the tanh table. When the calculation in LSTM is switched from sigmoid to tanh, the data that has been loaded can be used immediately, reducing the time required to load the lookup table.

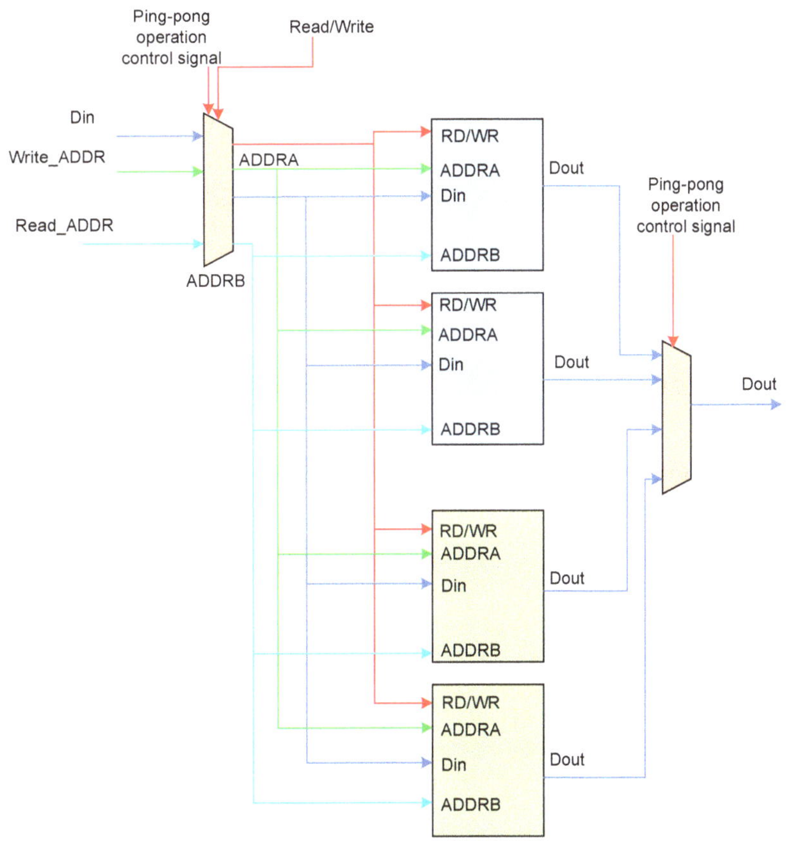

Figure 7. Structural diagram of the double buffer.

4. Instruction Design

In Section 4, we introduce the design details of the instruction set and the execution process of the instruction.

4.1. Information in the Instruction

In our design, the LSTM is split into a sequence of instructions. A matrix multiplication unit calculates eight data at the same time, so each instruction operates on eight data. When the eight data operations performed by one instruction are completed, the next instruction is read in, and the eight data operations of the next instruction are performed. When all instructions finish running, the calculation of the entire network also terminates at the same time. The information contained in an instruction is shown in Table 2, below.

The instruction contains the following information, information related to the input and output data, information related to the weights, information related to the post-processing parameters, information related to the activation functions, information related to matrix multiplication, information related to FIFO, and storage information about the output data. The functions of the instructions are abundant, which simplifies the design of the state machine of the control module in the hardware circuit. The hardware circuit will read the information from the instruction and decide which data to read for calculation and which address of the memory to store the calculated result, according to the information.

Table 2. Details in one instruction.

Field in the Instruction	Meaning
dat_input_len	length of the input data
dat_output_len	length of the output data
wgt_addr	address of weight
wgt_len	length of weight
post-processing_para_addr	address of post-processing parameter
load_sigmoid_cmd	whether to write the sigmoid/tanh parameter table
sigmoid_buffer_cmd	which buffer the sigmoid/tanh parameter table is written to
sigmoid_addr	address of the sigmoid/tanh parameter table
loop_num_of_mmu	the number of times the matrix multiplication unit calculates
vector_addr	address of the input vector
ct_addr_cmd	address of c_t
uram_store_addr_cmd	the address where the output is written to URAM
fifo2postprocess	whether the data in the last FIFO enters the post-processing unit
fifo2hbm	whether the data in the last FIFO is written to HBM

4.2. From LSTM to Instructions

With the network model and instruction definition, an LSTM can be split into a sequence of instructions. The instructions and atomic operations after LSTM decomposition are shown in Table 3. An atomic operation means an operation that can be completed by a matrix multiplication unit or a post-processing unit at one time.

The seven equations (Equations (1)–(7)) are broken down into 24 atomic operations. The 24 atomic operations are completed by a total of five instructions. Which atomic operations each instruction corresponds to, is also shown in Table 3.

Because matrix multiplication occupies the main amount of the calculation, we use the atomic operation of matrix multiplication as a separation point, to separate the data calculated by each instruction. The operations that can be completed by one instruction, include a matrix multiplication operation and several post-processing operations. For example, instruction 3 performs atomic operations from number 10 to number 16 while instruction 4 performs atomic operations from number 17 to number 23. When instruction 3 is executed, the matrix multiplication unit will perform atomic operation 10. After atomic operation 3 is completed, the post-processing unit will perform the calculation of atomic operation 11 to atomic operation 16, and at the same time, the matrix multiplication unit will perform operations on atomic operation 17. The matrix multiplication unit and the post-processing unit perform calculations simultaneously, which can achieve high throughput and performance. The total amount of instructions is related to the size of the matrix. If the output dimensions of the five matrix multiplications in LSTM are all 1024, and LSTM iterates 32 times, then the total number of instructions is $32 \times 1024 \times 5/8 = 20{,}480$.

4.3. Memory Reuse during Instruction Execution

During instruction execution, matrix multiplication and post-processing calculations require frequent reading and writing to memory. In order to reduce memory usage and maximize memory utilization, we have designed a memory reuse scheme.

For a matrix multiplication unit, it is relatively simple to read the data from the URAM and write it into the BRAM after calculation, and there is no need to consider memory reuse. For the post-processing unit, because the calculation it performs needs to read and write data repeatedly, we achieve the purpose of reducing storage resources by reusing BRAM, FIFO, and URAM.

Table 3. Splitting the LSTM into a sequence of instructions.

Number	Instruction Number	Atomic Operation
1	1	g_{mat_mult}=$W_g[X_t$:$Y_{t-1}]$
2	1	g_{mat_mult}=g_{mat_mult}+b_g
3	1	$g_t = \sigma(g_{mat\_mult})$
4	2	$i_{mat\_mult} = W_i[X_t : Y_{t-1}]$
5	2	$i_{mat\_mult} = i_{mat\_mult} + b_i$
6	2	$i_{elem\_mult} = W_{ic} \odot c_{t-1}$
7	2	$i_{elem\_add} = i_{mat\_mult} + i_{elem\_mult}$
8	2	$i_t = \sigma(i_{elem\_add})$
9	2	$g_t i_t = g_t \odot i_t$
10	3	$f_{mat\_mult} = W_f * [X_t : Y_{t-1}]$
11	3	$f_{mat\_mult} = f_{mat\_mult} + b_f$
12	3	$f_{elem\_mult} = W_{fc} \odot c_{t-1}$
13	3	$f_{elem\_add} = f_{mat\_mult} + f_{elem\_mult}$
14	3	$f_t = \sigma(f_{elem\_add})$
15	3	$f_t c_{t-1} = f_t \odot c_{t-1}$
16	3	$c_t = f_t c_{t-1} + g_t i_t$
17	4	$o_{mat\_mult} = W_o[X_t : Y_{t-1}]$
18	4	$o_{mat\_mult} = W_o[X_t : Y_{t-1}] + b_o$
19	4	$o_{elem\_mult} = W_{oc} \odot c_t$
20	4	$o_{elem\_add} = o_{mat\_mult} * o_{elem\_mult}$
21	4	$o_t = \sigma(o_{elem\_add})$
22	4	$tanh\_c_t = tanh(c_t)$
23	4	$m_t = o_t \odot tanh\_c_t$
24	5	$y_t = W_{ym} m_t$

Taking atomic operation 4 to atomic operation 9 in Table 3 as an example, the well-designed data flow is shown in Table 4.

Table 4. Memory reuse scheme.

Operations	Input Data1	Data1 Memory	Input Data2	Data2 Memory	Output Data3	Data3 Memory
Bias adding stage1	i_{mat_mult}	BRAM[35:0]	bias	post-processing parameter	i_{mat_mult}	BRAM[35:0]
Bias adding stage2	i_{mat_mult}	BRAM[35:0]	scale	post-processing parameter	i_{mat_mult}	BRAM[15:0]
Element-wise multiplication	W_{ic}	post-processing parameter	c_{t-1}	BRAM[63:48]	i_{elem_mult}	BRAM[31:16]
Element-wise addition	i_{elem_mult}	BRAM[31:16]	i_{mat_mult}	BRAM[15:0]	i_{elem_add}	BRAM[15:0]
Sigmoid	i_{elem_add}	BRAM[15:0]	none	none	i_t	BRAM[15:0]
Element-wise multiplication	i_t	BRAM[15:0]	g_t	FIFO	$g_t i_t$	FIFO

In order to maximize the use of storage resources, we put multiple data into the same address, by bit width division. As shown in Table 4, the result of adding bias is placed in BRAM[15:0], c_{t-1} is placed in BRAM [63:48], and the result of element-wise multiplication is placed in BRAM[31:16]. $g_t i_t$ is placed in the FIFO and will be read from the FIFO to participate in the calculation when it needs to be calculated. For other instructions, the operation is similar and will not be repeated here. Through the reuse of storage resources, the intermediate results in all LSTM calculations can be stored on only one BRAM, one URAM, and one FIFO.

5. Experimental Results

5.1. Experimental Setup

In our experiments, we implemented the LSTM network mentioned in Section 2. Firstly, instructions are generated according to the network structure and hardware structure. The required information, such as data address and length, is stored in the instruction, and the instruction is generated through a Python script. Our accelerator is implemented using Verilog code. Vivado 2020.1 and Vitis 2020.1 are our design development tools.

There are two SLRs (super logic region) on the Xilinx Alevo U50 accelerator card, and we deploy one kernel on each SLR. A kernel is composed of 32 cascaded computing units. Each computing unit consists of a matrix multiplication unit containing 64 DSPs and a post-processing unit containing 2 DSPs. So in our design, there are 64 computing units distributed on two SLRs, which means that the input data of 64 users can be calculated at the same time.

5.2. Resource Utilization

After placing and routing, the resources occupied by the accelerator are shown in Table 5. The resources used by the platform in Table 5, are the resources needed by Xilinx FPGA to communicate between the kernel and the host. Through the platform in the FPGA and the Xilinx Runtime (XRT) in the CPU, the host and the kernel can easily transmit data. It can be seen from Table 5 that 4224 DSPs are used, which is consistent with the result calculated according to the hardware structure. In order to store input data and intermediate results, the usage of URAM and BRAM in our design is within an acceptable range. The usage of BRAM is 282 through our resource reuse, otherwise more BRAM would be used. The kernel uses 122,935 lookup tables (LUTs), mainly because the instruction-driven design reduces the complexity of the hardware design. If the control module is not implemented by instructions, more LUTs will be required than in the current design. The usage of registers is slightly larger, mainly due to the cascaded design, which needs to register a lot of data.

Table 5. Resource usage and utilization.

Resource	LUT	LUT As MEM	Register	BRAM	URAM	DSP
Total	870,016	402,016	1,740,032	1344	640	5940
Used by platform	145,219	25,745	253,970	180	0	4
Used by kernel	122,935	5536	407,690	282	384	4224
Utilization (platform + kernel)	30.8%	7.8%	38.0%	34.4%	60.0%	71.2%

5.3. Performance Comparison

We compared our results with those of others. Since we have not seen work implementing LSTM using the same Alevo U50 card, we compare it with work using FPGAs with a similar amount of computational resources. The comparison results are in Table 6.

In our design, the overall circuit runs at 280 MHz. Our accelerator achieves a performance of 2036 GOPS at a 16-bit data bit width. The power consumption of our design, in Table 6, is obtained through Xilinx's power analysis tool. Compared with the work in [9], our design has higher performance and resource utilization, while using the same 16 bit data precision. Compared with the work in [16,18], our performance is lower but the data bit width used in [16,18] is 8 bit. Because there are rich int8 multipliers in Stratix 10 GX2800, int8 performance will be relatively high in Stratix 10 GX2800. Our design has higher resource utilization compared to [16,18]. Compared with [6,22], we obtain higher performance with higher data precision. Due to the simultaneous computation of data for 64 users, our design occupies a relatively large on-chip storage space (14.74 MB) and has a higher latency than other works. On the one hand, because the instructions in our design are executed continuously, the DSP in the matrix multiplication unit has almost no idle time and operates continuously. On the other hand, because the DSP of the matrix multiplication unit and the DSP of the post-processing unit work in parallel, and the DSP of the post-processing unit realizes different operations by configuring different modes, the hardware utilization of our work reached 86.1%, which exceeds the current designs.

Table 6. Comparison with previous work.

	C-LSTM [9]	**FCCM2020 [18]**	**Remarn [16]**	**SIBBS [22]**	**FDTT-LSTM [6]**	**Our Work**
Year	2018	2020	2022	2022	2023	
FPGA	Vertex-7	Stratix 10 GX 2800	Stratix 10 GX 2800	Kintex KU115	XCKU060	Alevo U50
Network	LSTM	LSTM	LSTM	LSTM	LSTM	LSTM
Frequency	200 MHz	260 MHz	260 MHz	200 MHz	200 MHz	280 MHz
Precision	16 bit	8 bit	8 bit	8 bit (weight) 12 bit (activation)	12 bit	16 bit
DSP used	2676 (74.3%)	4368 (76%)	4368 (76%)	4224 (76.52%)	972 (53%)	4224 (71%)
On-chip memory used (MB)	4.24	24.56	24.80	2.40	1.01	14.74
Performance (GOPS)	131.1	4790	6965	712.6	273.5	2036
Latency (ms)	0.0167	0.033	N/A	0.00104	N/A	9.786
Power	22 W	125 W	125 W	12.0 W	18.6 W	32.3 W
Power efficiency (GOPS/W)	6.0	38.32	55.72	59.3	14.7	62.84
LSTM hardware utilization	12.2%	56.1%	81.6%	42.2%	70.3%	86.1%

6. Conclusions

This paper presents an instruction-driven LSTM accelerator. The hardware part of the accelerator consists of a matrix multiplication unit and a post-processing unit. The matrix multiplication unit adopts a staggered reading scheme in the weight reading stage, to reduce the consumption of register resources. The post-processing unit completes operations such as element-wise multiplication, element-wise addition, batch normalization, and bias addition, by using only two DSPs, through resource sharing. Multi-layer LSTM and large LSTM can be decomposed into a series of instructions for execution, each of which executes a certain amount of data. Our design is implemented on the Xilinx Alevo U50 card, and the experimental results show that our design can achieve 2036 GOPS performance, and the resource utilization of hardware exceeds the existing designs. Our design currently uses 16-bit data and will support optimization of low-bit precision data in future work. Using low-bit data, such as 8-bit data, can further enhance the overall performance. Our research can be used in scenarios such as speech recognition and machine translation, in the data center.

Author Contributions: Conceptualization, N.M., H.Y. and Z.H.; methodology, N.M., H.Y. and Z.H.; software, N.M.; validation, N.M.; investigation, N.M., H.Y. and Z.H.; writing—original draft preparation, N.M.; writing—review and editing, N.M., H.Y. and Z.H.; supervision, H.Y.; project administration, H.Y.; funding acquisition, H.Y. All authors have read and agreed to the published version of the manuscript.

Funding: This work was supported by the National Natural Science Foundation of China, under grant 61876172.

Conflicts of Interest: The authors declare no conflict of interest.

References

1. Amodei, D.; Ananthanarayanan, S.; Anubhai, R.; Bai, J.; Battenberg, E.; Case, C.; Casper, J.; Catanzaro, B.; Cheng, Q.; Chen, G.; et al. Deep Speech 2: End-to-End Speech Recognition in English and Mandarin. In Proceedings of the 33rd International Conference on Machine Learning, New York, NY, USA, 19–24 June 2016; pp. 173–182.
2. Jouppi, N.P.; Hyun Yoon, D.; Ashcraft, M.; Gottscho, M.; Jablin, T.B.; Kurian, G.; Laudon, J.; Li, S.; Ma, P.; Ma, X.; et al. Ten Lessons From Three Generations Shaped Google's TPUv4i: Industrial Product. In Proceedings of the 2021 ACM/IEEE 48th Annual International Symposium on Computer Architecture (ISCA), Valencia, Spain, 14–18 June 2021; pp. 1–14.
3. Rybalkin, V.; Sudarshan, C.; Weis, C.; Lappas, J.; Wehn, N.; Cheng, L. Efficient Hardware Architectures for 1D- and MD-LSTM Networks. *J. Signal Process. Syst.* **2020**, *92*, 1219–1245. [CrossRef]
4. Que, Z.; Zhu, Y.; Fan, H.; Meng, J.; Niu, X.; Luk, W. Mapping Large LSTMs to FPGAs with Weight Reuse. *J. Signal Process. Syst.* **2020**, *92*, 965–979. [CrossRef]
5. Azari, E.; Vrudhula, S. An Energy-Efficient Reconfigurable LSTM Accelerator for Natural Language Processing. In Proceedings of the 2019 IEEE International Conference on Big Data (Big Data), Los Angeles, CA, USA, 9–12 December 2019; pp. 4450–4459.
6. Liu, M.; Yin, M.; Han, K.; DeMara, R.F.; Yuan, B.; Bai, Y. Algorithm and hardware co-design co-optimization framework for LSTM accelerator using quantized fully decomposed tensor train. *Internet Things* **2023**, *22*, 100680. [CrossRef]
7. Que, Z.; Nakahara, H.; Nurvitadhi, E.; Boutros, A.; Fan, H.; Zeng, C.; Meng, J.; Tsoi, K.H.; Niu, X.; Luk, W. Recurrent Neural Networks With Column-Wise Matrix–Vector Multiplication on FPGAs. *IEEE Trans. Very Large Scale Integr. (VLSI) Syst.* **2022**, *30*, 227–237. [CrossRef]
8. Que, Z.; Wang, E.; Marikar, U.; Moreno, E.; Ngadiuba, J.; Javed, H.; Borzyszkowski, B.; Aarrestad, T.; Loncar, V.; Summers, S.; et al. Accelerating Recurrent Neural Networks for Gravitational Wave Experiments. In Proceedings of the 2021 IEEE 32nd International Conference on Application-specific Systems, Architectures and Processors (ASAP), Piscataway, NJ, USA, 7–9 July 2021; pp. 117–124.
9. Wang, S.; Li, Z.; Ding, C.; Yuan, B.; Qiu, Q.; Wang, Y.; Liang, Y. C-LSTM: Enabling Efficient LSTM using Structured Compression Techniques on FPGAs. In Proceedings of the 2018 ACM/SIGDA International Symposium on Field-Programmable Gate Arrays, Monterey, CA, USA, 25–27 February 2018; pp. 11–20.
10. Azari, E.; Vrudhula, S. ELSA: A Throughput-Optimized Design of an LSTM Accelerator for Energy-Constrained Devices. ACM Trans. Embed. *Comput. Syst.* **2020**, *19*, 3.
11. Cao, S.; Zhang, C.; Yao, Z.; Xiao, W.; Nie, L.; Zhan, D.; Liu, Y.; Wu, M.; Zhang, L. Efficient and Effective Sparse LSTM on FPGA with Bank-Balanced Sparsity. In Proceedings of the 2019 ACM/SIGDA International Symposium on Field-Programmable Gate Arrays, Seaside, CA, USA, 24–26 February 2019; pp. 63–72.
12. Chen, J.; Hong, S.; He, W.; Moon, J.; Jun, S.-W. Eciton: Very Low-Power LSTM Neural Network Accelerator for Predictive Maintenance at the Edge. In Proceedings of the 2021 31st International Conference on Field-Programmable Logic and Applications (FPL), Dresden, Germany, 30 August–3 September 2021; pp. 1–8.
13. Ioannou, L.; Fahmy, S.A. Streaming Overlay Architecture for Lightweight LSTM Computation on FPGA SoCs. *ACM Trans. Reconfigurable Technol. Syst.* **2022**, *16*, 8. [CrossRef]
14. Kim, T.; Ahn, D.; Lee, D.; Kim, J.-J. V-LSTM: An Efficient LSTM Accelerator using Fixed Nonzero-Ratio Viterbi-Based Pruning. *IEEE Trans. Comput.-Aided Des. Integr. Circuits Syst.* **2023**, 1. [CrossRef]
15. Nurvitadhi, E.; Kwon, D.; Jafari, A.; Boutros, A.; Sim, J.; Tomson, P.; Sumbul, H.; Chen, G.; Knag, P.; Kumar, R.; et al. Why Compete When You Can Work Together: FPGA-ASIC Integration for Persistent RNNs. In Proceedings of the 2019 IEEE 27th Annual International Symposium on Field-Programmable Custom Computing Machines (FCCM), San Diego, CA, USA, 28 April–1 May 2019; pp. 199–207.

16. Que, Z.; Nakahara, H.; Fan, H.; Li, H.; Meng, J.; Tsoi, K.H.; Niu, X.; Nurvitadhi, E.; Luk, W. Remarn: A Reconfigurable Multi-threaded Multi-core Accelerator for Recurrent Neural Networks. *ACM Trans. Reconfigurable Technol. Syst.* **2022**, *16*, 4. [CrossRef]
17. Que, Z.; Nakahara, H.; Fan, H.; Meng, J.; Tsoi, K.H.; Niu, X.; Nurvitadhi, E.; Luk, W. A Reconfigurable Multithreaded Accelerator for Recurrent Neural Networks. In Proceedings of the 2020 International Conference on Field-Programmable Technology (ICFPT), Maui, HI, USA, 9–11 December 2020; pp. 20–28.
18. Que, Z.; Nakahara, H.; Nurvitadhi, E.; Fan, H.; Zeng, C.; Meng, J.; Niu, X.; Luk, W. Optimizing Reconfigurable Recurrent Neural Networks. In Proceedings of the 2020 IEEE 28th Annual International Symposium on Field-Programmable Custom Computing Machines (FCCM), Fayetteville, AR, USA, 3–6 May 2020; pp. 10–18.
19. Que, Z.; Nugent, T.; Liu, S.; Tian, L.; Niu, X.; Zhu, Y.; Luk, W. Efficient Weight Reuse for Large LSTMs. In Proceedings of the 2019 IEEE 30th International Conference on Application-specific Systems, Architectures and Processors (ASAP), New York, NY, USA, 15–17 July 2019; pp. 17–24.
20. Rybalkin, V.; Ney, J.; Tekleyohannes, M.K.; Wehn, N. When Massive GPU Parallelism Ain't Enough: A Novel Hardware Architecture of 2D-LSTM Neural Network. *ACM Trans. Reconfigurable Technol. Syst.* **2021**, *15*, 2.
21. Rybalkin, V.; Pappalardo, A.; Ghaffar, M.M.; Gambardella, G.; Wehn, N.; Blott, M. FINN-L: Library Extensions and Design Trade-Off Analysis for Variable Precision LSTM Networks on FPGAs. In Proceedings of the 2018 28th International Conference on Field Programmable Logic and Applications (FPL), Dublin, Ireland, 27–31 August 2018; pp. 89–96.
22. Jiang, J.; Xiao, T.; Xu, J.; Wen, D.; Gao, L.; Dou, Y. A low-latency LSTM accelerator using balanced sparsity based on FPGA. Microprocess. *Microsystems* **2022**, *89*, 104417. [CrossRef]
23. He, D.; He, J.; Liu, J.; Yang, J.; Yan, Q.; Yang, Y. An FPGA-Based LSTM Acceleration Engine for Deep Learning Frameworks. *Electronics* **2021**, *10*, 681. [CrossRef]
24. Hochreiter, S.; Schmidhuber, J. Long Short-Term Memory. *Neural Comput.* **1997**, *9*, 1735–1780. [CrossRef] [PubMed]
25. Sak, H.; Senior, A.; Françoise, B. Long short-term memory recurrent neural network architectures for large scale acoustic modeling. In Proceedings of the 15th Annual Conference of the International Speech Communication Association, Singapore, 14–18 September 2014; pp. 338–342.

Disclaimer/Publisher's Note: The statements, opinions and data contained in all publications are solely those of the individual author(s) and contributor(s) and not of MDPI and/or the editor(s). MDPI and/or the editor(s) disclaim responsibility for any injury to people or property resulting from any ideas, methods, instructions or products referred to in the content.

electronics

MDPI

Article

A New FPGA-Based Task Scheduler for Real-Time Systems

Lukáš Kohútka [1,*] and Ján Mach [2]

[1] Institute of Informatics, Information Systems and Software Engineering, Slovak University of Technology in Bratislava, 812 43 Bratislava, Slovakia
[2] Institute of Computer Engineering and Applied Informatics, Slovak University of Technology in Bratislava, 812 43 Bratislava, Slovakia
* Correspondence: lukas.kohutka@stuba.sk

Abstract: This research demonstrates a novel design of an FPGA-implemented task scheduler for real-time systems that supports both aperiodic and periodic tasks. The periodic tasks are automatically restarted once their period has expired without any need for software intervention. The proposed scheduler utilizes the Earliest-Deadline First (EDF) algorithm and is optimized for multi-core CPUs, capable of executing up to four threads simultaneously. The scheduler also provides support for task suspension, resumption, and enabling inter-task synchronization. The design is based on priority queues, which play a crucial role in decision making and time management. Thanks to the hardware acceleration of the scheduler and the hardware implementation of priority queues, it operates in only two clock cycles, regardless of the number of tasks in the system. The results of the FPGA synthesis, performed on an Intel FPGA device (Cyclone V family), are presented in the paper. The proposed solution was validated through a simplified version of the Universal Verification Methodology (UVM) with millions of test instructions and random deadline and period values.

Keywords: real-time; task scheduling; EDF; FPGA; hardware acceleration; periodic tasks; CPU; SoC

Citation: Kohútka, L.; Mach, J. A New FPGA-Based Task Scheduler for Real-Time Systems. *Electronics* **2023**, *12*, 1870. https://doi.org/10.3390/electronics12081870

Academic Editors: Andres Upegui, Andrea Guerrieri and Laurent Gantel

Received: 28 February 2023
Revised: 8 April 2023
Accepted: 12 April 2023
Published: 15 April 2023

Copyright: © 2023 by the authors. Licensee MDPI, Basel, Switzerland. This article is an open access article distributed under the terms and conditions of the Creative Commons Attribution (CC BY) license (https://creativecommons.org/licenses/by/4.0/).

1. Introduction

Real-time systems are a type of embedded system that handles tasks that require real-time processing. The success of these tasks is dependent on both the accuracy of the results and the time at which they are finished. Missing deadlines for real-time tasks can be considered a failure, just like calculating the wrong results. Therefore, real-time system reliability is achieved when tasks are finished within the specified time frame [1,2].

Task scheduling algorithms often use priority queues that are performed within software. Such software-based solutions work well for relatively simple and tiny real-time systems containing a limited number of tasks. However, as the number and complexity of tasks increase, the performance and constant response time become more critical, especially for safety-critical systems. Meeting the deadlines of tasks is considered a reliability requirement, as missing a deadline is seen as a failure of the system. Even using a high-performance processor does not guarantee that all tasks will meet their deadlines, which is why a dedicated task scheduler is necessary for real-time and safety-critical systems [3–7].

Real-time task scheduling is a critical aspect of many computing systems, particularly those that are safety-critical, time-critical, or both. Unfortunately, software-based solutions for task scheduling have several limitations, particularly with regards to the time consumed for scheduling, the determinism, dependability, and predictability of the system, and the algorithms used for scheduling. In an ideal world, a task scheduler for real-time systems would be able to generate an optimal schedule, using zero CPU time to perform the scheduling itself. This would allow all of the CPU time to be used for executing the tasks themselves rather than for scheduling. Of course, this ideal scenario is unlikely to be achievable in practice, and real-world schedulers will always spend CPU time to some extent. However, it is possible to minimize the time spent on scheduling and make it as constant as possible, which would improve the determinism, dependability, and predictability of a real-time system. One drawback of task schedulers implemented in software is that they are restricted in the algorithms they can use, as they need to consume a very low and constant quantity of processor time. This often results in the use of priority-based scheduling algorithms, rather than algorithms based on task deadlines, leading to a lack of robustness and efficiency in the scheduling process. One possible solution to these limitations is the use of hardware acceleration for task scheduling. By implementing the task scheduler in hardware, it is possible to use deadline-based scheduling algorithms that minimize the time spent on scheduling and make it constant as well. This would allow for a more robust and efficient scheduling process, which would be a significant improvement over software-based solutions [7–16].

Hardware-accelerated task schedulers have been traditionally used for simple systems that consist solely of aperiodic hard real-time (RT) tasks. However, these solutions have proven to be insufficient for more complex and robust real-time systems that have a higher number of tasks and a greater variety of task types. To address this issue, a more advanced and sophisticated task scheduler is required. One that has the ability to support a diverse range of processes/threads. Apart from aperiodic RT tasks, hard RT systems also perform periodic tasks. These tasks can be managed by task schedulers in the same way as aperiodic tasks. However, including dedicated support for periodic tasks directly within the HW-based scheduler can significantly improve the overall performance of the system. This is because the addition of this support in HW eliminates the need for any further software extensions for periodic task management because periodic tasks are autonomously rescheduled without using the CPU whenever a period is completed. As a result, the system can operate more efficiently and effectively, providing the desired level of real-time performance and reliability. Overall, the implementation of a more robust and complex task scheduler is necessary to meet the demands of modern real-time systems that have a greater number of tasks and a wider range of task types. This will ensure that the system can perform at its optimal level and deliver the desired level of performance, reliability, and functionality [17–24].

The aim of the research presented in this article is to design a new version of a coprocessor unit that is capable of scheduling tasks based on the Earliest-Deadline First (EDF) algorithm [7]. The EDF algorithm is widely regarded as a dynamic version of deadline-based scheduling algorithms [25], as it eliminates the need for assigning individual task priorities. One of the main challenges faced by modern CPUs is the parallel execution of multiple tasks, which is a result of the widespread adoption of the many-core paradigm in processor design. This poses significant complications for hardware-accelerated task scheduling, particularly for RT systems that contain periodic hard RT tasks and require inter-task synchronization. In order to address these challenges, the research focuses on ensuring that the coprocessor unit is highly efficient and scalable in terms of performance while also ensuring that the overall system remains reliable and deterministic. This is a critical aspect that must be considered when developing hardware-accelerated task scheduling systems for real-time systems. In principle, the research presented in this article seeks to provide a solution to the obstacles and challenges faced by modern CPUs in the implementation of hardware-accelerated task scheduling. By focusing on the implementation of the EDF algorithm and addressing the critical aspects of efficiency, scalability, reliability, and determinism, the proposed coprocessor unit is designed to deliver improved performance and functionality for real-time systems containing periodic hard RT tasks.

This paper is structured the following way: The paper's Section 2 covers task schedulers for real-time systems. In Section 3, the paper introduces a new solution for task scheduling. The proposed solution is verified in Section 4. The paper presents synthesis results for the proposed solution in Section 5 and discusses the outcomes. Section 6 contains an evaluation of the performance achieved by the proposed solution, including a comparison with software-based scheduling. Finally, the paper concludes with a summary in Section 7.

2. Related Work

There are a lot of scheduling algorithms, each with pros and cons [26,27]. A deep comparison of global, partitioned, and clustered EDF scheduling algorithms in software has been presented in [28]. The authors performed experiments on a 24-core Intel Xeon L7455 system, where each core was running at 2.13 GHz. For these algorithms, they analyzed the overheads of the algorithms for the scheduling of tasks, their release, context switching, and several other types of overheads. The outcome was that as the number of tasks increases, the scheduling overhead is increasing, mainly for the global EDF, where the worst-case scheduling overhead for 250 tasks was 200 µs. The partitioned and clustered EDF algorithms reached up to around 30 µs overhead.

The scheduling overhead analysis of several algorithms has also been done in [29]. The authors run tests on a single-core ATmega2560 clocked at 16 MHz. The overhead of two tested non-preemptive EDF schedulers increased with the number of tasks. The maximum scheduling overhead by the basic EDF for twelve tasks was 136 µs, while the Critical-Window EDF had a maximum overhead of 404 µs.

Authors in [30] used the profiler of the Virtual Machine for a comparison of two algorithms to get information about their scheduling overhead on multiple cores. The first algorithm (LRE-LT) tried very hard to ensure that all deadlines were met. On the other hand, the second algorithm (USG) tried to minimize task preemption and migrations between cores, so it was expected that the second one missed a few deadlines. It was shown that the scheduler has been invoked a lot more often in the LRE-L and that the decision-making process took longer. The result was that the time spent on scheduling was approximately 10,000 times longer in the LRE-LT than in the USG.

Task scheduling plays a crucial role in operating systems, as it determines which task (i.e., thread or process) should be running in the processor and in what order. The algorithms used for scheduling greatly impact these decisions. Classic operating systems typically schedule tasks based on their priorities of tasks, while RT systems must schedule tasks based on their deadlines. This is because meeting the deadlines of all hard real-time tasks is of the utmost importance in real-time systems. The Earliest-Deadline First (EDF) algorithm is one of the most widely used and well-known algorithms for scheduling hard real-time tasks. It operates by sorting all tasks based on their deadlines, with the task having the earliest deadline being selected for execution first. Since tasks need to be sorted according to their deadlines, priority queues are the ideal data structure for implementing the EDF algorithm. Also, task scheduling is a core and critical component of operating systems that must be carefully designed and implemented [7,31,32].

The ideal real-time task scheduler is one that schedules tasks optimally, ensuring that all tasks are completed before their deadlines while minimizing the overhead on the CPU. The more CPU time that is consumed by the scheduling algorithm, the less effective the CPU becomes at executing the scheduled tasks. It is inevitable that some CPU time will be consumed, as the scheduler must use at minimum one clock cycle for transferring data to and from the scheduling unit. However, to achieve optimal performance, the goal is to spend the minimum amount of processor time. Maintaining a predictable and deterministic embedded system is critical, and this requires that a constant amount of CPU time be spent on scheduling, regardless of the actual amount of tasks currently present in the scheduler or even the maximum amount of tasks that can be handled by the system (i.e., the capacity of the task queue). This ensures that the system operates in a consistent and predictable manner, making it easier to debug and optimize. These qualities are essential for ensuring that real-time systems operate reliably and efficiently.

Our previous research [33] resulted in the development of a novel real-time task scheduler based on the Earliest-Deadline First (EDF) algorithm. This scheduler was implemented as a coprocessor unit, and a comparison was made between HW and SW realizations with regards to efficiency and performance. The coprocessor is designed to consume two cycles of the processor's clock domain, no matter how many tasks the system contains. Subsequently, an improved variant of the coprocessor was developed to be suitable for dual-core CPUs. To solve the issue of conflicting situations where more than one CPU core attempts to access the coprocessor at the same time, two approaches were proposed and compared [34]. Finally, support for scheduling non-real-time tasks was added to the scheduler, utilizing priorities instead of deadlines [35].

In addition to our coprocessor, there are other existing solutions for HW-accelerated task scheduling on RT systems. Some of them are also utilizing the EDF algorithm [21,22,36,37]. One solution, presented in [37], uses the EDF algorithm as well, but with a limited maximum number of tasks of 64. On the other hand, another solution relies on task priorities instead of deadlines, which is not suitable for hard RT systems [38]. There are also other approaches that adopt a priority-based or static scheduling method [39–42]. One solution, presented in [43], supports EDF, LST (Least-Slack-Time) and priority-based scheduling. Schedulers based on genetic algorithms and fuzzy logic are presented in [44,45].

Our previous coprocessor solution is efficient for simple RT systems with hard RT tasks in conjunction with single-core CPUs. On the other hand, as RT systems become more complex and require higher performance, multi-core CPUs are often used, requiring a more complex task scheduler to support multiple cores. A thorough analysis of the suitability of the EDF algorithm in uniform multiprocessor systems reported that this algorithm is applicable in such cases too [37]. Some of the existing solutions also incorporate a method to monitor the remaining execution time of real-time tasks and predict potential deadline misses [38,41,42]. Based on the analysis of existing schedulers, we have decided to design a new RT task scheduler implemented as a coprocessor unit suitable for quad-core RT embedded systems and that would support periodic tasks and inter-task synchronization too.

The performance of real-time task schedulers based on deadlines relies heavily on the ability to sort tasks using their deadlines. To achieve this, a priority queue data structure is used as the central component in hardware-implemented task schedulers. There have been numerous designs for data sorting in priority queues for real-time systems, including the FIFO approach [36,38,40,46], Shift Registers [37,47–49], and Systolic Array [41,42,49,50]. Each of these architectures has been developed to provide efficient sorting capabilities, making them popular choices for use in real-time task scheduling.

The FIFO approach is extremely inefficient in terms of chip area and suitable for a small range of possible values (deadlines in this case), up to four or five bits only [36,38,40,46].

The architecture called Shift Registers is made up of homogeneous cells that each consist of a comparator, control logic, and registers to store one item. The cells are connected in a line, and each cell can exchange items with its two neighbors. The cells receive instructions simultaneously from the queue input, which results in an increase in the critical path length with an increase in the number of cells. The critical path length is a result of the bus width used for simultaneous instruction delivery and the exchange of control signals between all cells. The critical path issue of Shift Registers can be resolved by using a register at the inputs of cells, dedicating one clock cycle for the shared bus. The throughput of the Shift Registers architecture is one instruction per clock cycle. An illustration of a four-cell Shift Registers architecture can be seen in Figure 1 [37,47–49].

The architecture called Systolic Array is quite like Shift Registers, but it overcomes the critical path length issue by utilizing pipelining. The Systolic Array features homogenous cells, which are connected in a linear manner, with each cell having one neighboring cell to the left and right, excluding the first and last cell of the structure. The first cell in the queue serves as the only source of output for the entire queue and is receiving instructions through the queue's input. All instructions are gradually passed from one cell to the next cell, at a rate of one cell per clock cycle, in a manner similar to how instructions are processed through pipeline stages in pipelined processors. The throughput, however, of this architecture is smaller than Shift Registers because each delete instruction takes two clock cycles instead of just one. An example of the Systolic Array architecture consisting of four cells is displayed in Figure 2. The first cell on the right side contains the first item in the priority queue and serves as the interface between the surrounding circuits. Clock and reset are the only parallel signals [41,42,49,50].

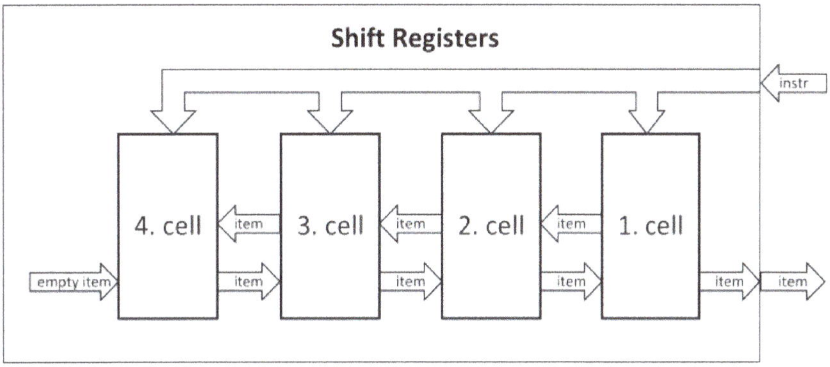

Figure 1. Shift Registers architecture example [48].

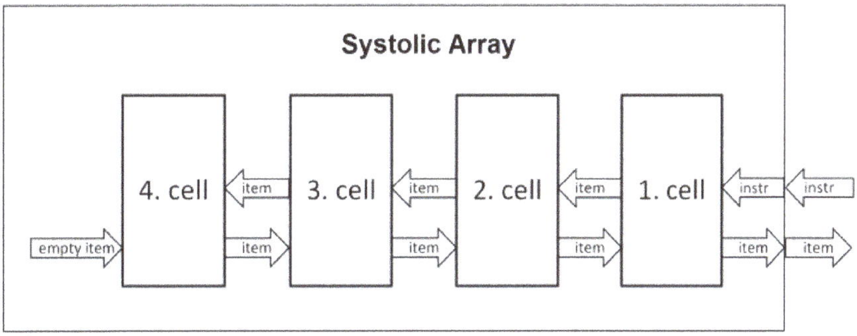

Figure 2. Systolic Array architecture example [41,42,49,50].

The Systolic Array priority queue brings a solution to the critical path issue present in Shift Registers. The instructions are propagated through the cells, with each cell representing one pipeline stage with pipeline registers. This means N clock cycles are required for an instruction to propagate via the entire queue, where N is the queue capacity. Since every cell is performing a different instruction at a time, the throughput of this structure is one instruction per clock cycle. However, after deleting an instruction, a pause (NOP) is needed for one cycle, resulting in a throughput of one instruction per two clock cycles. The priority queue is providing an updated output in just two cycles. Thus, response time is two clock cycles. This response time remains constant, regardless of the number of cells in the queue. In addition to the lower throughput of this architecture in comparison to Shift Registers, the second disadvantage is the almost doubled amount of flip flops needed for implementation of the Systolic Array [41,42,49,50].

3. Proposed Solution

The proposed solution is a task scheduler based on FPGA technology, which is designed as a coprocessor that receives instructions from the processor and sends back decisions about which instructions are supposed to be executed at the moment. This implies that the scheduler will be encapsulated or integrated into an existing CPU, much like any other coprocessor, for example a multiplier or divider. Figure 3 illustrates the architecture at the top level of abstraction of the top-level module of the designed coprocessor, which is comprised of seven submodules: Ready Queue, Waiting Queue, Idle Queue, Tasks Memory, Running Tasks, Control Unit, and Semaphore.

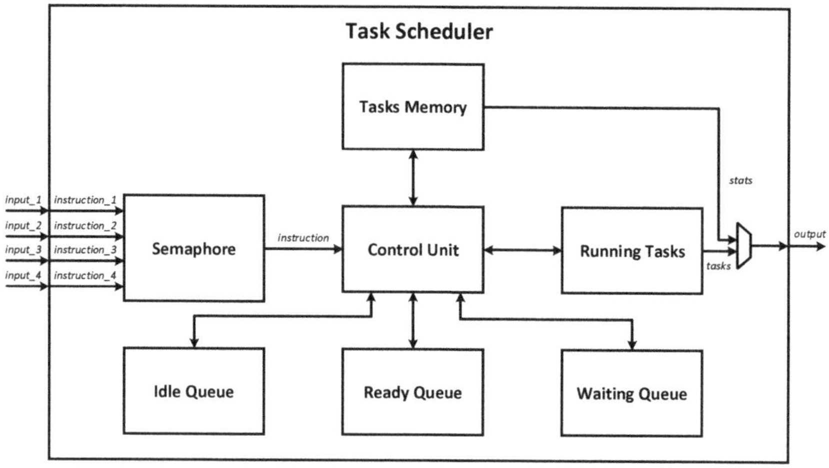

Figure 3. Block diagram of proposed task scheduler.

The scheduler top-level module contains four input ports, known as instr_1, instr_2, instr_3 and instr_4, which are used by the CPU to provide the coprocessor instructions for the scheduler. It is assumed that up to four tasks/processes/threads can run simultaneously (for example, on a quad-core CPU). The output port of the scheduler is providing valuable data for the processor. Using this output port, the processor also gets the information about which (up to four) processes are currently dedicated to run right now, and the processor is also able to obtain memory data from the Tasks Memory submodule this way as well.

The proposed task scheduler is providing the following list of new instructions:

- MEMORY_WRITE—can be used to create a new task in Tasks Memory or to modify already created tasks. This instruction performs a standard write operation into the memory.
- MEMORY_READ—can be used to read any information about tasks stored in Tasks Memory. This instruction performs a standard read operation from the memory.
- SCHEDULE_TASK—is used to schedule an existing task to be executed by the CPU. This causes the task to be moved to Running Tasks or Ready Queue, which is a decision based on the scheduling algorithm (i.e., deadline values). Task states that are stored inside Tasks Memory will be modified if needed, too.
- KILL_TASKL—is used to deschedule (i.e., to kill) an already scheduled task. As a result, the task is removed from the queues, such as Running Tasks, and the task state is set to IDLE_TASK in Tasks Memory.
- BLOCK_TASK—is used to temporarily block a scheduled task, forbidding its execution for a limited time. As a result, the state of the selected task is changed to WAITING, and the task is moved into the Waiting Queue. The task is blocked for a specified time only; therefore, a waiting time is set too. When the waiting time elapses, the blocked task will be automatically unblocked.
- UNBLOCK_TASK—is used to unblock a blocked task. As a result, an existing blocked task is released (i.e., unblocked), changing its state from WAITING to a different state, and this task is also removed from the Waiting Queue, returning the task back to the Ready Queue or Running Tasks. Since blocked tasks are automatically unblocked after a specific waiting time elapses, this instruction is just meant for unblocking the task earlier, eliminating the need to wait until the waiting time has elapsed.
- GET_RUNNING_TASKS—is used to obtain the list of running tasks or the task that is selected for execution in a particular processor core using the scheduler output port. This information is provided by the Running Tasks module.

3.1. Ready Queue

The Ready Queue is a key component in the task scheduler, as it holds all tasks that are ready for execution but are waiting their turn. This component is designed as a priority queue, sorting tasks that are ready by their deadline values so that the next task to be executed can be quickly identified. The sorting of tasks is performed by utilizing the Shift Registers architecture, which was explained in Section 2. This architecture is composed of sorting cells, each of which consists of a comparator for comparing deadlines, control logic for deciding when and what to store in this cell, and a register as an actual storage element to remember the ID of the task and its deadline. These cells are able to move tasks to neighboring cells, and they receive instructions simultaneously via a common bus. Ready Queue ensures that the tasks are executed in a timely manner, based on their deadlines, and provides the necessary information to the CPU for task selection and execution.

3.2. Waiting Queue

The Waiting Queue is an integral part of the task scheduler and is designed to hold all the temporarily suspended or blocked tasks. This component is also implemented as a priority queue, just like the Ready Queue, to sort the waiting tasks based on their remaining waiting times. The waiting tasks are referred to as such because they are temporarily put on hold and are waiting to be unblocked. This can occur in two ways: either through the execution of the UNBLOCK_TASK instruction or if the remaining time for waiting has elapsed. Once a task is unblocked, it is extracted from the Waiting Queue. This prioritization of waiting tasks ensures that the task scheduler can effectively manage the execution of multiple tasks and maintain a high level of performance.

The Waiting Queue is used only for the inter-task synchronization instructions BLOCK_TASK and UNBLOCK_TASK. While this queue is allowing users to block and unblock tasks, the inter-task synchronization logic itself is not implemented and is only supported by providing these two instructions. It is up to the software extension to decide whether and when a particular task is supposed to be blocked, for how long it is blocked, and eventually, whether a blocked task is unblocked before the block time elapses.

3.3. Idle Queue

The Idle Queue is a module designed to hold idle periodic tasks (i.e., state = idle), either because they were completed naturally or terminated using the KILL_TASK instruction. This component is only relevant for periodic tasks; tasks that are not periodic are not stored in the Idle Queue. The module is structured as yet another priority queue, similar to Waiting Queue and Ready Queue, with each task being sorted based on their remaining period times. The output of Idle Queue represents the next periodic task that is going to finish its period. When current time reaches the task's period time, the task is extracted from the Idle Queue and rescheduled to start a new instance of this task for the new period. It is efficient handling of periodic tasks, and it is possible despite the fact that only one task can be extracted from the Idle Queue at a time. If more tasks end their period at the same time, i.e., they need rescheduling simultaneously, then those tasks that are rescheduled later automatically adjust their remaining deadline time by the delay incurred. Thus, such tasks may be rescheduled in any sequence under the condition that the rescheduled tasks get adjusted for their remaining deadlines appropriately.

Unlike the Waiting Queue, the Idle Queue does not need any instructions to be called from the CPU in order to manage the idle (completed) periodic tasks that are waiting for their next period. Whenever the period of the periodic task elapses, the task is automatically moved from the Idle Queue back to Running Tasks. The state of this task is automatically changed from idle to ready or running (depending on the EDF logic, i.e., deadlines of tasks). This automation is provided by the Control Unit module.

3.4. Semaphore

The Semaphore component is designed to handle conflicts that arise when multiple CPU cores simultaneously attempt to use the scheduler, e.g., to schedule a new task or to kill a task in exactly the same clock cycle. This situation is referred to as a conflict. To resolve conflicts, the Semaphore is a module that is responsible for arbitrating instructions by selecting one of the instructions as the arbitration winner and the rest of the instructions becoming losers. The winner's instruction is passed on to the Control Unit. The loser instructions cannot be executed immediately; therefore, other CPU cores are stalled. Semaphore module uses a widely known algorithm called Round-Robin, which guarantees that the arbitration is fair, and the load is evenly balanced. This is an important aspect of the FPGA-based task scheduler design, as it ensures that all CPU cores have equal access to the scheduler coprocessor and that no core is favored over the others.

The total number of possible conflicts that can arise in the system is eleven, and each conflict is identified by a different combination of CPU cores attempting to use the scheduler at the same time. When two processor cores try to access the coprocessor simultaneously, there are six combinations possible: 1_2 (processor core 1 and core 2 conflict), 1_3 (core 1 and core 3 conflict), 1_4 (core 1 and core 4 conflict), 2_3 (core 2 and core 3 conflict), 2_4 (core 2 and core 4 conflict), and 3_4 (core 3 and core 4 conflict). In the case where three CPU cores attempt to access the scheduler at the same time, there are four possible combinations: 1_2_3 (core 1, core 2, and core 3 conflict), 1_2_4 (core 1, core 2, and core 4 conflict), 1_3_4 (core 1, core 3, and core 4 conflict), and 2_3_4 (core 2, core 3, and core 4 conflict). The final combination occurs if all four processor cores try to access the coprocessor simultaneously, and this is named 1_2_3_4.

The Semaphore module has two essential demands, a primary and a secondary one. The most critical demand is to have a limited maximum number of delays caused by conflicts and to keep this number low. This demand is vital since the scheduler is designed for RT embedded systems. The second demand is to have fairness among all CPU cores, ensuring that each core has a similar chance of winning the conflict and accessing the scheduler immediately.

The design of the Semaphore module is based on a 2-bit counter to represent four distinct states. These states, referred to as 1234, 2143, 3412, and 4321, are used to determine the priority order in case of any of these eleven possible combinations of access conflicts. For instance, if the current state is 1234, then processor core 1 is prioritized over core 2, processor core 2 is prioritized over core 3, and processor core 3 is prioritized over core 4. To resolve the conflicts, the current state is shifted to the new state through an increment of a 2-bit counter. The decision was taken to limit the amount of possible priority order permutations to just four instead of the original twenty-four permutations in order to simplify the FSM that is deciding which processor core wins the arbitration. This results in a relatively simple design of the FSM consisting of four states instead of twenty-four states, reducing chip area and energy consumption. These four states have been carefully selected to ensure symmetry, fairness, and rotations after each conflict. Whenever there is a conflict, the order is updated by shifting FSM to the next state. The 2-bit counter is used to represent the states in the following manner:

- Counter set to "00" is representing state 1234, and is incremented to "01".
- Counter set to "01" is representing state 2143, and is incremented to "10".
- Counter set to "10" is representing state 3412, and is incremented to "11".
- Counter set to "11" is representing state 4321, and is incremented to "00".

The scenarios in which two or more CPU cores conflict, along with the winner selection, are presented in Table 1. This table outlines the different potential scenarios for conflicting CPU cores, with each line representing a possible conflict. The columns represent the four possible orders, where one order is used at the moment based on which state the FSM is currently in. It is noticeable that the primary as well as secondary demands for the Semaphore were satisfied, as each processor core stalled three times in a row, at most, and the distribution of wins among the CPU cores is even. The rotating nature of the states/orders ensures that the worst-case scenario for one instruction to be completed is $2N$ clock cycles (in the event that all processor cores attempt to access the coprocessor continuously), where N is the number of processor cores (i.e., $N = 4$). On the other hand, the best-case access time is just two cycles. Consequently, the amount of time it takes for one instruction to be executed on a quad-core CPU ranges from two to eight clock cycles, which depends on the frequency of access conflicts.

Table 1. Table of conflict resolutions in Semaphore module.

Title 1	1234	2143	3412	4321
1_2	1	2	1	2
1_3	1	1	3	3
1_4	1	1	4	4
2_3	2	2	3	3
2_4	2	2	4	4
3_4	3	4	3	4
1_2_3	1	2	3	3
1_2_4	1	2	4	4
1_3_4	1	1	3	4
2_3_4	2	2	3	4
1_2_3_4	1	2	3	4

A block diagram of the Semaphore module is shown in Figure 4, based on the previously described information. The module includes 3 multi-bit multiplexers, 2 D-FFs, 5 AND gates, 4 NAND gates, 2 inverters, a Winner Selector submodule, and a Conflict Detector. The Winner Selector and Conflict Detector submodules only require the instruction's bit, which indicates if the instruction is valid or not (if the processor is attempting to access the coprocessor). The Winner Selector submodule selects the winner of arbitration, i.e., deciding which of the valid instructions should be selected. The selection signals are called SEL00, SEL01, and SEL1. These signals drive the control inputs of MUXes that pass the winning instruction to an output port named "instr". This port is used by Control Unit module.

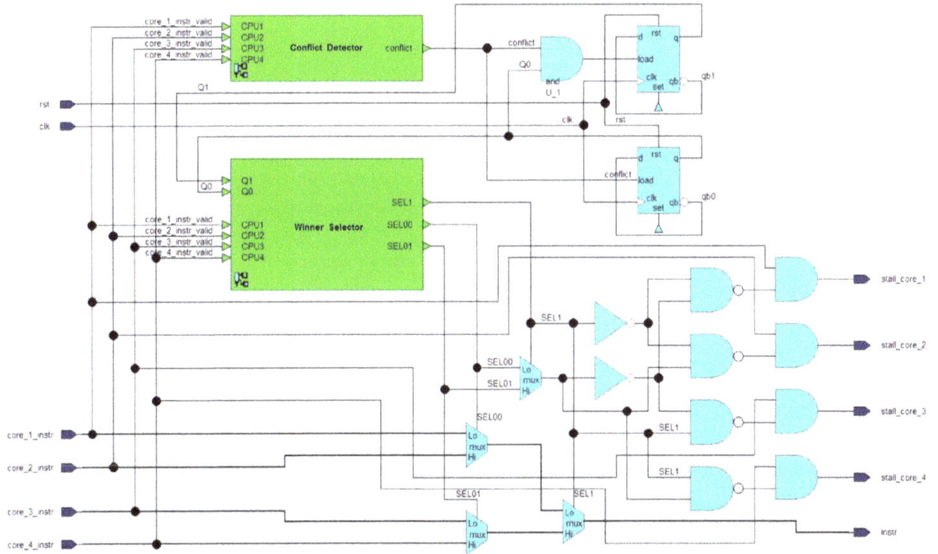

Figure 4. Block diagram of Semaphore module.

The Semaphore module in Figure 4 also outputs four single-bit outputs to the individual processor cores, named stall_core_#. These signals serve to notify the respective processor core when the attempt to access the coprocessor is declined due to losing an arbitration caused by the situation when another core is trying to use the coprocessor as well (i.e., during a conflict). The processor core receiving the stall response must wait until the stall is active. In the meantime, it can perform other operations. The signals core_#_instr_valid and stall_core_# act as a means of handshaking-based communication between the processor and coprocessor.

The Conflict Detector submodule is illustrated in Figure 5 as a logic circuit. It consists of one 6-input NAND gate and six basic 2-input NAND gates. The purpose of this submodule is to detect whether 2 or more processor cores are trying to access the coprocessor simultaneously. If two or more inputs are set to 1, the conflict output will be 1 to indicate a conflict. However, if there is only one valid instruction from a single CPU core at most, the conflict output will be 0.

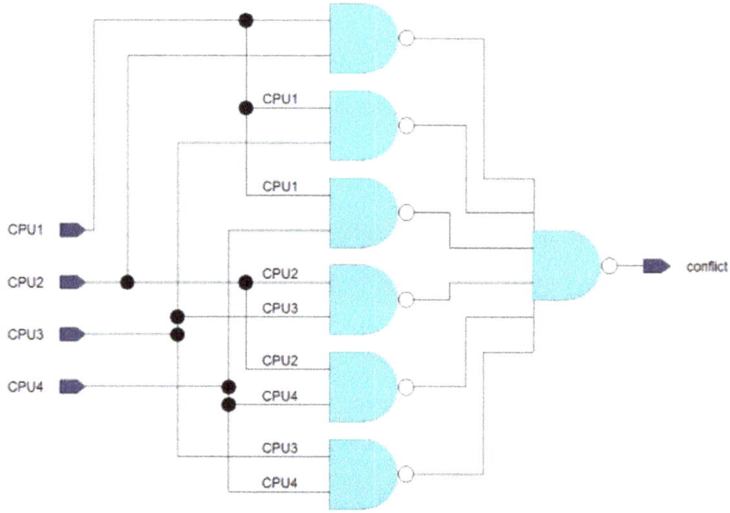

Figure 5. Logic circuit of Conflict Detector.

The Winner Selector submodule, as depicted in Figure 6, implements its decision logic through the use of one 3-input NOR and three 2-input NORs for the SEL1 output, and two 2-input NANDs for each of the SEL00 and SEL01 outputs. The circuit includes two inverter gates for creating inverted input signals as well. The decision logic for this module is the same as that outlined in Table 1. The Q1 and Q0 inputs represent the actual state of the FSM. The inputs CPU1, CPU2, CPU3, and CPU4 represent information about which processor cores are attempting to access the coprocessor at the moment. The output signals SEL00, SEL01, and SEL1 are needed for controlling the multiplexing presented in Figure 4.

3.5. Running Tasks

The Running Tasks module holds the tasks that are currently being executed. With a capacity of four tasks, it can execute up to four tasks at once. To minimize unnecessary task switches, this component can assign a task that was previously running on one CPU core to another CPU core due to task preemption. By doing so, the number of task switches is reduced to the minimum possible amount, and the scheduling overhead is minimized this way.

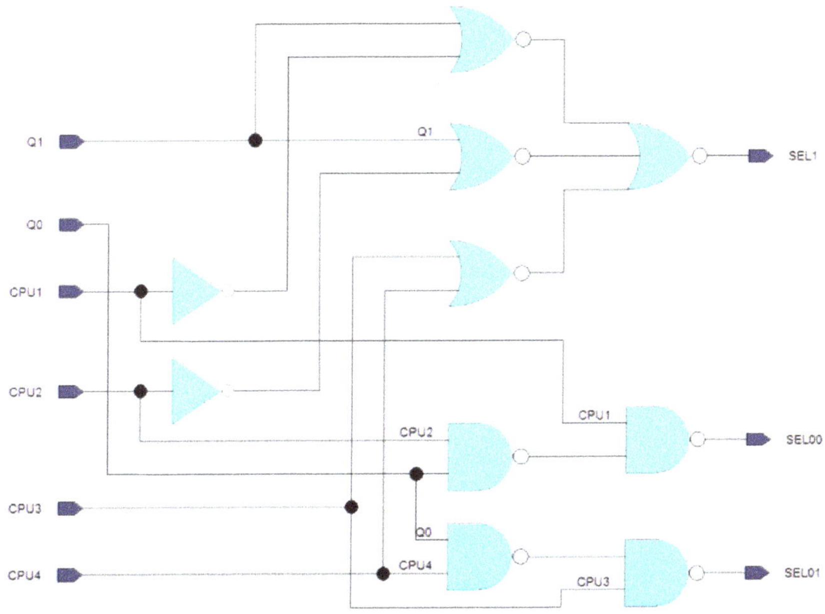

Figure 6. Winner Selector logic circuit.

The control logic of Running Tasks performs decisions about whether to keep the current tasks or make changes. If a task is terminated, the task that has the lowest deadline value within the Ready Tasks submodule is moved into the Running Tasks submodule. When the coprocessor is scheduling new tasks, a preemption may occur, depending on the task deadline and the deadlines of currently running tasks. Whenever the task is scheduled with an earlier deadline than the deadline of any running task, the running task with the latest deadline is replaced, causing task preemption, and execution of the preempted task is suspended. If preemption occurs, the preempted task is stored in the Ready Tasks module; otherwise, the new task is added to it.

The Running Tasks component has five comparators and requires two clock cycles for decision logic due to the length of the critical path in the combination logic. The results of the comparison performed in the first cycle are stored in registers holding temporary results along with 2 bits for task identification. The first cycle performs two parallel comparisons, comparing running_task_core_1 with running_task_core_2 and running_task_core_3 with running_task_core_4. During the second cycle, the deadlines of temporary results obtained from the registers are compared, which may result in a preemption, replacing one running task with a new one. The previously running task that is being replaced is sent to the Ready Queue if preemption occurs. Regardless of preemption, one of the tasks is sent to the Ready Tasks module—either the new task (i.e., no preemption occurs) or one of the previously running tasks (i.e., preemption occurs).

3.6. Control Unit

The Control Unit component manages all task queue modules (Idle Queue, Waiting Queue, Ready Queue, and Running Tasks) and reads from and writes to the Tasks Memory module. When a valid instruction is received from the Semaphore component, the Control Unit module decodes it and performs it by providing data and control signals to surrounding modules. The Control Unit directly controls all task queue components, except for the Ready Queue, which it can only access indirectly using the Running Tasks module. When no valid instruction is received from the CPU or Semaphore, the Control Unit can still transfer tasks autonomously between the Running Tasks module and Idle Queue module and between the Running Tasks module and Waiting Queue module. This ensures that managing waiting or blocked tasks, including periodic idle tasks, is done automatically, reducing the CPU time needed for task scheduling, which is crucial for RT tasks.

Control Unit is also responsible for managing states of tasks. There are four possible states the task can have in this scheduler: IDLE, RUNNING, READY and WAITING. The IDLE state is used for tasks that are not yet scheduled or are completed already. This is especially important for periodic tasks to determine that the task is finished, and that the scheduler is waiting for the next period of the task to automatically schedule a new instance of the periodic task. The RUNNING state is used for tasks that are running, i.e., being executed at the moment. The READY state is used for tasks that are scheduled and ready to be executed but have not yet been chosen for execution due to other tasks being prioritized over the ready task. WAITING state is used for those tasks that are blocked by the BLOCK_TASK instruction, so these tasks are waiting to be unblocked either by the UNBLOCK_TASK instruction or by elapsing the waiting time set by the BLOCK_TASK instruction. These states and possible changes between the states are depicted in Figure 7.

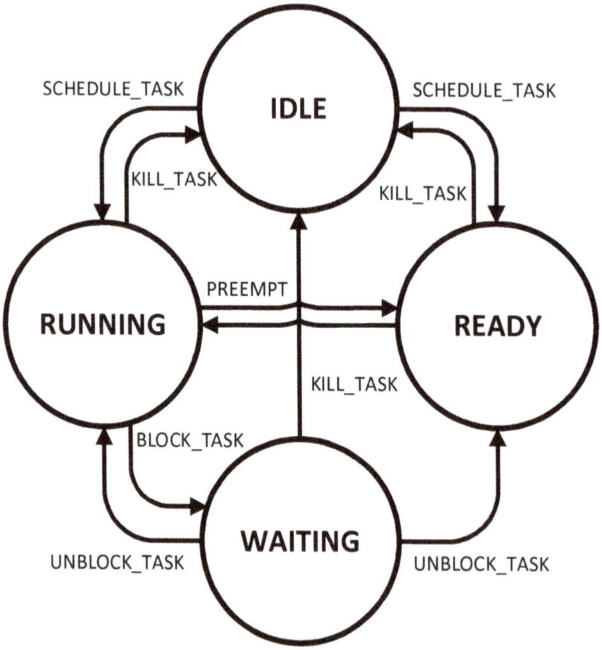

Figure 7. State diagram of task states.

3.7. Tasks Memory

The Tasks Memory component is a multi-port standard memory designed to support various features in other components. Though it could potentially be implemented using SRAM, it has been realized with registers due to the need for multiple read/write ports. While this implementation based on registers consumes more chip area, the added read/write ports are essential to the functions described in other components.

The Tasks Memory stores all information about tasks, including task type, task state, ID of parent task, and timing characteristics such as starting/remaining deadline, starting/remaining period (if the task is periodic), and starting/remaining execution time of the task. While the starting timing characteristics are provided by the CPU when a task is created, the remaining timing characteristics are automatically maintained by the scheduler itself. The memory's layout is outlined in Table 2, with the lowest three bits of address being reserved to choose specific data within the particular task while the upper bits are utilized to choose a task.

Table 2. Memory map of Tasks Memory.

Address Bits 2 Downto 0	Field	Number of Bits
000	ID of parent task	8
001	Task state + task type	4 + 5
010	Remaining deadline time	20
011	Remaining period time	20
100	Remaining execution time	20
101	Starting deadline time	20
110	Starting period time	20
111	Starting execution time	20

4. Design Verification

The task scheduling coprocessor that was introduced was described using the SystemVerilog language and then tested through simulations. The ModelSim tool was used to perform these simulations. Additionally, to the SystemVerilog language, a simplified variant of the Universal Verification Methodology (UVM) was also utilized during the verification phase. The interface of the scheduler was quite simple, making it possible to simplify the use of UVM as well. In this scenario, a single transaction within the UVM test equates to one instruction executed in two clock cycles, eliminating the need for agents to interface with the device under test (DUT).

The verification phase involved the use of one test procedure that generated constrained random input values, a scoreboard, and a predictor. This test generated millions of instructions with deterministic instruction opcodes and unique task ID values, but with random timing values. The predictor is responsible for predicting the expected output of DUT (Design Under Test) based on the input values. The predictor behaves similarly to the DUT but at a higher level of abstraction, similar to high-level software languages. The predictor's description was purely sequential and high-level, utilizing the SystemVerilog priority queue data structure and the sort() procedure to order the tasks within each queue. Figure 8 demonstrates the testbench that was applied for verification simulations.

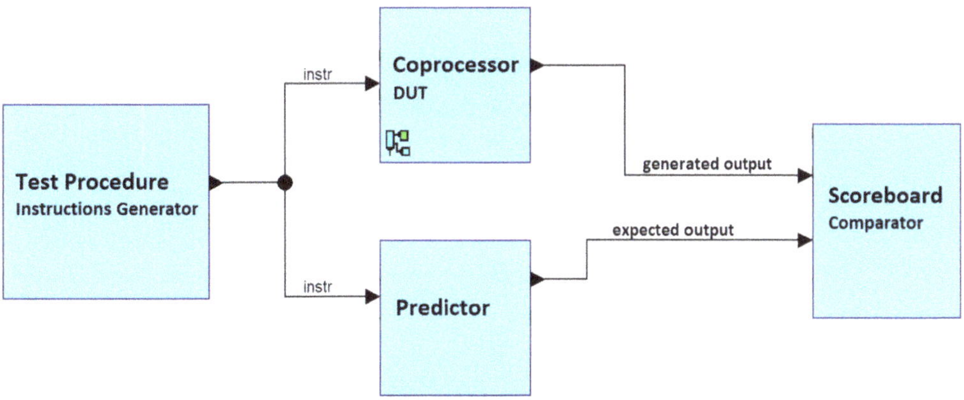

Figure 8. Testbench architecture.

To verify that the designed scheduler is working properly and as expected, more than 2,000,000 iterations of the test were performed, each containing at least 1000 randomly generated instructions. The full capacity of the scheduler was utilized during this test. Scheduler parameters were set to the following values during the design verification:

eight bits for task IDs, a capacity of Ready Queue set to sixty tasks, and twenty bits for random deadlines, execution times, and task periods.

The verification process was thorough, ensuring that the proposed task scheduler was functioning as expected. The use of the SystemVerilog language and UVM, along with the test procedure and predictor, provided a comprehensive and efficient method for verifying the coprocessor unit's behavior. The results of these simulations demonstrate the reliability and effectiveness of the task scheduler, making it a suitable solution for real-time task management.

5. Synthesis Results

The proposed task scheduler was implemented on an Intel Cyclone V FPGA, more specifically the 5CSEBA6U23I7 device. The synthesis process was performed using the Intel Quartus Prime 16.1 Lite Edition tool. To ensure that the scheduler would operate properly, a static timing analysis was performed to determine the maximum clock frequency for each version of the design.

The synthesis results presented in Table 3 indicate that the maximum clock frequency of all versions of the proposed scheduler is 105 MHz or higher. The critical path, i.e., the path that is limiting the maximum clock frequency, was found in priority queues. Therefore, increasing the size of these priority queues has an impact on the maximum clock frequency (fMax). However, the resource requirements, as measured by Adaptive Logic Module (ALM) consumption, are relatively low considering the large capacity of current FPGA devices, which often have hundreds of thousands, if not millions, of ALMs.

Table 3. FPGA synthesis results.

Tasks Capacity	ALMs	Registers	fMax (MHz)
8	334	325	177.99
16	591	541	156.98
24	832	756	137.85
32	1067	976	127.67
40	1324	1181	122.05
48	1576	1403	118.10
56	1817	1624	107.49
64	2044	1833	105.52

It is important to note that the Tasks Capacity, or the maximum number of tasks the scheduler can handle, has a direct and significant impact on both the resource costs and the maximum clock frequency. As the Tasks Capacity increases, the logic utilization increases and the timing performance (fMax) decreases. The implementation of each timing variable, such as deadline, period, waiting time, and execution time, uses twenty bits, while the task ID is comprised of eight bits.

6. Performance Evaluation

This section demonstrates the performance benefits of using the proposed task scheduler instead of the existing software-based task scheduler, the G-EDF (Global Earliest Deadline First) algorithm that was presented in [28] and used on a 24-core Intel Xeon CPU running at 2.13 GHz.

Two use cases of the proposed scheduler are considered: one case when a CPU that is running four tasks in parallel (i.e., four CPU cores) is used, and the second case when a CPU with one task (i.e., one CPU core) is used. In both cases, the proposed scheduler is running on the FPGA described in the previous section at 100 MHz. Using this clock frequency means that one clock cycle takes ten nanoseconds, which is equal to 0.01 us. Table 4 shows the worst-case CPU overhead of task scheduling, i.e., the time needed to schedule one task (to call one SCHEDULE_TASK instruction). This overhead is displayed in microseconds (us). The four CPU cores version takes seven clock cycles in the worst-case scenario, which occurs when the CPU core has to wait for six clock cycles plus one clock cycle for calling the SCHEDULE_TASK instruction.

Table 4. Worst-case CPU overhead of task scheduling comparison in microseconds (us).

Number of Tasks	G-EDF on Xeon [28]	Proposed Scheduler (with 4 CPU Cores)	Proposed Scheduler (with 1 CPU Core)
25	20	0.07	0.01
36	28	0.07	0.01
50	42	0.07	0.01
64	51	0.07	0.01
100	140	0.07	0.01

The worst-case overhead of software-based EDF scheduling is around 20 us when 25 tasks are used. If this scheduling is hardware-accelerated using the proposed solution, then the overhead drops to less than 0.1 us, effectively reducing the CPU overhead more than 200-times, i.e., more than 99.5% overhead reduction is achieved. It is worth noting that the overhead of the proposed solution is constant with respect to the number of tasks—unlike in software-based scheduling, where the CPU overhead increases with the number of tasks. Thus, the relative reduction of CPU overhead is even 99.95% (i.e., 2000-times lower overhead) when 100 tasks are used. Therefore, the proposed HW-based scheduler has much better scalability for the growing number of tasks, allowing more complex real-time systems with a higher number of tasks to be implemented.

The proposed solution significantly outperformed the existing software-based scheduling despite the fact that the existing solution used a 2.13 GHz clock and the proposed solution used only a 100 MHz clock. If the proposed solution was implemented using cutting-edge ASIC technology together with a CPU, then the performance benefits would be even bigger, reducing the scheduling overhead down to around 3.5 ns (i.e., 0.0035 us) and further reducing the overhead by around 200-times if a 2 GHz clock was used. On the other hand, since the main limitation of the proposed solution is the amount of HW resources (i.e., chip area in ASIC or Look-Up Tables in FPGA), which heavily depends on the number of tasks to be supported, the FPGA technology brings a significant advantage in the form of configurability and reconfigurability (including the partial reconfiguration feature of FPGAs) of the scheduler. In FPGA, the scheduler can be configured to have optimal task capacity, whereas in ASIC, the scheduler cannot be reconfigured for optimal task capacity, which can lead to significant waste of chip area if the actual real-time system is using much less tasks than the scheduler capacity. The optimal setting of scheduler capacity is hard to determine as it depends on the real-time system requirements and needs.

The overall performance benefits of using a hardware-accelerated task scheduler also depend on what percentage of the CPU time is spent on the scheduling and what percentage is used for the actual execution of scheduled tasks. This depends on the actual application of the real-time system and the granularity of tasks—whether the application uses a few big tasks or many smaller ones. However, regardless of the actual application, by accelerating the task scheduling in hardware, it is possible to use almost all of the CPU time for the actual execution of scheduled tasks instead of scheduling those tasks. Thanks to Moore's Law, the costs of hardware-accelerated scheduling are gradually lower and lower, causing the overall benefits to outweigh the costs.

7. Conclusions

The proposed task scheduler is a novel solution that implements the Earliest-Deadline First (EDF) scheduling algorithm on an FPGA. This scheduler is well-suited for complex real-time systems that consist of a mixture of aperiodic hard RT tasks, periodic hard RT tasks, and non-real-time (best-effort) tasks. It leverages a priority queue-based approach to handle all types of tasks efficiently and with ease.

The scheduler uses priority queues not only to sort the ready tasks but also to handle idle periodic tasks and waiting/blocked tasks. As a result, managing these types of tasks is straightforward, allowing for an autonomous handling process with no need for software extension, with the only exception being the Tasks Memory initialization. Additionally, the priority queue-based approach makes the proposed scheduling solution highly efficient in managing periodic RT tasks and readily extensible to include task synchronization and inter-task communication capabilities.

This scheduler is optimized for use on quad-core CPUs, which can execute up to four tasks in parallel. All of the supported instructions take a maximum of three clock cycles to complete, no matter the system configuration or the actual or maximum amount of tasks in the system, provided there are not any conflicts between multiple processor cores attempting to access the scheduler simultaneously. However, in the event of such conflicts, an extra delay of two to six cycles may occur, leading to a maximum latency of nine clock cycles per instruction in the worst-case scenario.

In conclusion, the proposed task scheduler is a highly efficient solution for real-time systems that can handle a diverse range of tasks, including aperiodic hard RT tasks, periodic hard RT tasks, and non-real-time (best-effort) tasks. Its utilization of priority queues simplifies the handling of periodic idle tasks and blocked/waiting tasks, making it a suitable option for systems that require minimal software intervention. The scheduler's performance is further enhanced by its compatibility with quad-core CPUs and its ability to execute instructions in a few cycles, regardless of the number of tasks the system contains. These features make the proposed task scheduler an attractive option for complex real-time systems that require efficient task management and optimal performance.

Author Contributions: Conceptualization, L.K.; methodology, L.K.; software, L.K.; validation, L.K.; formal analysis, L.K.; investigation, L.K.; resources, L.K. and J.M.; data curation, L.K.; writing—original draft preparation, L.K.; writing—review and editing, L.K. and J.M.; visualization, L.K.; supervision, L.K.; project administration, L.K.; funding acquisition, L.K. All authors have read and agreed to the published version of the manuscript.

Funding: The work reported here was supported by the Operational Programme Integrated Infrastructure for the project Advancing University Capacity and Competence in Research, Development and Innovation (ACCORD) (ITMS code: 313021X329), co-funded by the European Regional Development Fund (ERDF). This publication was also supported in part by the Slovak national project KEGA 025STU-4/2022, APVV-19-0401 and APVV-20-0346.

Data Availability Statement: Not applicable.

Conflicts of Interest: The authors declare no conflict of interest.

References

1. Mall, R. *Real-Time Systems: Theory and Practice*, 2nd ed.; Pearson Education India: Delhi, India, 2008; ISBN 978-81-317-0069-3.
2. O'Reilly, C.A.; Cromarty, A.S. *"Fast" Is Not "Real-Time" in Designing Effective Real-Time AI Systems*; Application of Artificial Intelligence II; SPIE: New York, NY, USA, 1985; Volumes 5–8, pp. 249–257. [CrossRef]
3. Stankovic, J.A.; Ramamritham, K. *Tutorial Hard Real-Time Systems*; Computer Society Press: Washington, DC, USA, 1988.
4. Buttazzo, G.; Stankovic, J. Adding Robustness in Dynamic Preemptive Scheduling. In *Responsive Computer Systems: Steps toward Fault-Tolerant Real-Time Systems*; Fussell, D.S., Malek, M., Eds.; Kluwer Academic Publishers: Boston, MA, USA, 1995.
5. Caccamo, M.; Buttazzo, G. Optimal scheduling for fault-tolerant and firm real-time systems. In Proceedings of the Fifth International Conference on Real-Time Computing Systems and Applications (Cat. No.98EX236), Hiroshima, Japan, 27–29 October 1998; pp. 223–231.
6. Buttazzo, G.C.; Sensini, F. Optimal deadline assignment for scheduling soft aperiodic tasks in hard real-time environments. *IEEE Trans. Comput.* **1999**, *48*, 1035–1052. [CrossRef]
7. Buttazzo, G.; Conticelli, F.; Lamastra, G.; Lipari, G. Robot control in hard real-time environment. In Proceedings of the Fourth International Workshop on Real-Time Computing Systems and Applications, Taipei, Taiwan, 27–29 October 1997; pp. 152–159.
8. Buttazzo, G.C. *Hard Real-Time Computing Systems: Predictable Scheduling Algorithms and Applications*; Springer: New York, NY, USA, 2011. [CrossRef]
9. Spuri, M.; Buttazzo, G.; Sensini, F. Robust aperiodic scheduling under dynamic priority systems. In Proceedings of the 16th IEEE Real-Time Systems Symposium, Pisa, Italy, 5–7 December 1995; pp. 210–219.

10. Heath, S. *Embedded Systems Design*; Newnes: Newton, MA, USA, 2003; ISBN 0750655461.
11. Lee, I.; Leung, J.Y.-T.; Son, S.H. *Handbook of Real-Time and Embedded Systems*; Chapman & Hall/CRC: Boca Raton, FL, USA, 2007; ISBN 9781584886785.
12. Joseph, M. *Real-Time Systems Specification, Verification and Analysis*; Prentice Hall International: London, UK, 2001.
13. Marwedel, P. *Embedded System Design: Embedded Systems Foundations of Cyber-Physical Systems*; Springer: Berlin/Heidelberg, Germany, 2010; ISBN 9400702566.
14. Pohronská, M. Utilization of FPGAs in Real-Time and Embedded Systems. In *Proceedings of the Informatics and Information Technologies Student Research Conference*; Bratislava, Slovakia, 29.4.2009; Bielikova, M., Ed.; Vydavateľstvo STU: Bratislava, Slovakia, 2009.
15. Lange, A.B.; Andersen, K.H.; Schultz, U.P.; Sorensen, A.S. HartOS—A Hardware Implemented RTOS for Hard Real-time Applications. *FAC Proc. Vol.* **2012**, *45*, 207–213. [CrossRef]
16. Liu, S.; Ding, Y.; Zhu, G.; Li, Y. Hardware scheduler of Real-time Operating. *Adv. Sci. Technol. Lett.* **2013**, *31*, 159–160.
17. Gergeleit, M.; Becker, L.B.; Nett, E. Robust scheduling in team-robotics. In Proceedings of the International Parallel and Distributed Processing Symposium, Nice, France, 22–26 April 2003; p. 8.
18. Norollah, A.; Derafshi, D.; Beitollahi, H.; Fazeli, M. RTHS: A Low-Cost High-Performance Real-Time Hardware Sorter, Using a Multidimensional Sorting Algorithm. *IEEE Trans. Very Large Scale Integr. (VLSI) Syst.* **2019**, *27*, 1601–1613. [CrossRef]
19. Derafshi, D.; Norollah, A.; Khosroanjam, M.; Beitollahi, H. HRHS: A High-Performance Real-Time Hardware Scheduler. *IEEE Trans. Parallel Distrib. Syst.* **2020**, *31*, 897–908. [CrossRef]
20. Suman, C.; Kumar, G. Performance Enhancement of Real Time System using Dynamic Scheduling Algorithms. In Proceedings of the 2019 IEEE 5th International Conference for Convergence in Technology (I2CT), Bombay, India, 29–31 March 2019; pp. 1–6.
21. Teraiya, J.; Shah, A. Optimized scheduling algorithm for soft Real-Time System using particle swarm optimization technique. *Evol. Intell.* **2021**, *15*, 1935–1945. [CrossRef]
22. Norollah, A.; Kazemi, Z.; Sayadi, N.; Beitollahi, H.; Fazeli, M.; Hely, D. Efficient Scheduling of Dependent Tasks in Many-Core Real-Time System Using a Hardware Scheduler. In Proceedings of the 2021 IEEE High Performance Extreme Computing Conference (HPEC), Waltham, MA, USA, 20–24 September 2021; pp. 1–7. [CrossRef]
23. Kotaba, O.; Nowotsch, J.; Paulitsch, M.; Petters, S.M.; Theiling, H. Multicore in Real-Time Systems—Temporal Isolation Challenges Due to Shared Resources. Available online: http://www.cister-labs.pt/docs/1044 (accessed on 24 February 2023).
24. Wandeler, E.; Maxiaguine, A.; Thiele, L. Quantitative Characterization of Event Streams in Analysis of Hard Real-Time Applications. In *Real-Time Systems*; Springer: Berlin/Heidelberg, Germany, 2005; Volume 29, pp. 205–225. [CrossRef]
25. Liu, C.L.; Layland, J.W. Scheduling Algorithms for Multiprogramming in a Hard-Real-Time Environment. *J. ACM* **1973**, *20*, 46–61. [CrossRef]
26. Omar, H.K.; Jihad, K.H.; Hussein, S.F. Comparative analysis of the essential cpu scheduling algorithms. *Bull. Electr. Eng. Inform.* **2021**, *10*, 2742–2750. [CrossRef]
27. Ramesh, P.; Ramachandraiah, U. Performance evaluation of real time scheduling algorithms for multiprocessor systems. In Proceedings of the 2015 International Conference on Robotics, Automation, Control and Embedded Systems (RACE), Chennai, India, 18–20 February 2015; pp. 1–4. [CrossRef]
28. Bastoni, A.; Brandenburg, B.B.; Anderson, J.H. An Empirical Comparison of Global, Partitioned, and Clustered Multiprocessor EDF Schedulers. In Proceedings of the 2010 31st IEEE Real-Time Systems Symposium, San Diego, CA, USA, 30 November–3 December 2010; pp. 14–24. [CrossRef]
29. Nasri, M.; Davis, R.I.; Brandenburg, B.B. FIFO with Offsets: High Schedulability with Low Overheads. In Proceedings of the 2018 IEEE Real-Time and Embedded Technology and Applications Symposium (RTAS), Porto, Portugal, 11–13 April 2018; pp. 271–282. [CrossRef]
30. Alhussian, H.; Zakaria, N.; Abdulkadir, S.J.; Fageeri, S.O. Performance evaluation of real-time multiprocessor scheduling algorithms. In Proceedings of the 2016 3rd International Conference on Computer and Information Sciences (ICCOINS), Kuala Lumpur, Malaysia, 15–17 August 2016; pp. 310–315. [CrossRef]
31. Churnetski, K. Real-Time Scheduling Algorithms, Task Visualization. Ph.D. Thesis, Computer Science Department Rochester Institute of Technology, Rochester, NY, USA, 2006.
32. Mohammadi, A.; Akl, S.G. *Scheduling Algorithms for Real-Time Systems*; School of Computing Queens University: Kingston, ON, Canada, 2005.
33. Kohutka, L.; Vojtko, M.; Krajcovic, T. Hardware Accelerated Scheduling in Real-Time Systems. In Proceedings of the Engineering of Computer Based Systems Eastern European Regional Conference, Brno, Czech Republic, 27–28 August 2015; pp. 142–143. [CrossRef]
34. Kohutka, L.; Stopjakova, V. Task scheduler for dual-core real-time systems. In Proceedings of the 23rd International Conference Mixed Design of Integrated Circuits and Systems, Lodz, Poland, 23–25 June 2016; pp. 474–479. [CrossRef]
35. Kohútka, L.; Stopjaková, V. Improved Task Scheduler for Dual-Core Real-Time Systems. In Proceedings of the 2016 Euromicro Conference on Digital System Design (DSD), Limassol, Cyprus, 31 August–2 September 2016; pp. 471–478. [CrossRef]
36. Bloom, G.; Parmer, G.; Narahari, B.; Simha, R. *Real-Time Scheduling with Hardware Data Structures*; IEEE Real-Time Systems Symposium: San Juan, PR, USA, 2010.

37. Tang, Y.; Bergmann, N.W. A Hardware Scheduler Based on Task Queues for FPGA-Based Embedded Real-Time Systems. *IEEE Trans. Comput.* **2015**, *64*, 1254–1267. [CrossRef]
38. Starner, J.; Adomat, J.; Furunas, J.; Lindh, L. Real-Time Scheduling Co-Processor in Hardware for Single and Multiprocessor Systems. In Proceedings of the EUROMICRO Conference, Prague, Czech Republic, 2–5 September 1996; pp. 509–512. [CrossRef]
39. Varela, M.; Cayssials, R.; Ferro, E.; Boemo, E. Real-time scheduling coprocessor for NIOS II processor. In Proceedings of the VIII Southern Conference Programmable Logic, Bento Gonçalves, Brazil, 20–23 March 2012; pp. 1–6. [CrossRef]
40. Ferreira, C.; Oliveira, A.S.R. Hardware Co-Processor for the OReK Real-Time Executive. *Eletrón. Telecomun.* **2010**, *5*, 160–166.
41. Ong, S.E.; Lee, S.C. SEOS: Hardware Implementation of Real-Time Operating System for Adaptability. In Proceedings of the 2013 First International Symposium on Computing and Networking, Matsuyama, Japan, 4–6 December 2013; pp. 612–616. [CrossRef]
42. Kim, K.; Kim, D.; Park, C.H. Real-Time Scheduling in Heterogeneous Dual-core Architectures. In Proceedings of the 12th International Conference on Parallel and Distributed Systems, Minneapolis, MN, USA, 12–15 July 2006. [CrossRef]
43. Wulf, C.; Willig, M.; Goehringer, D. RTOS-Supported Low Power Scheduling of Periodic Hardware Tasks in Flash-Based FPGAs. *Microprocess. Microsyst.* **2022**, *26*, 104566. [CrossRef]
44. Slimani, K.; Hadaoui, R.; Lalam, M. Hardware Fuzzy Scheduler for Real-Time Independent Tasks. *J. Circuits Syst. Comput.* **2022**, *31*, 2250155. [CrossRef]
45. Khare, A.; Patil, C.; Chattopadhyay, S. Task mapping and flow priority assignment of real-time industrial applications for network-on-chip based design. *Microprocess. Microsyst.* **2020**, *77*, 103175. [CrossRef]
46. Ferreira, C.M.; Oliveira, A.S. RTOS Hardware Coprocessor Implementation in VHDL. In Proceedings of the Intellectual Property/Embedded Systems Conference, Boston, MA, USA, 21–24 September 2009.
47. Chandra, R.; Sinnen, O. Improving Application Performance with Hardware Data Structures. In Proceedings of the 2010 IEEE International Symposium on Parallel & Distributed Processing, Workshops and Phd Forum (IPDPSW), Atlanta, GA, USA, 19–23 April 2010; pp. 1–4. [CrossRef]
48. Moon, S.W. Scalable Hardware Priority Queue Architectures for High-Speed Packet Switches. In Proceedings of the Third IEEE Real-Time Technology and Applications Symposium, Montreal, QC, Canada, 9–11 June 1997; pp. 203–212. [CrossRef]
49. Kohútka, L. Efficiency of Priority Queue Architectures in FPGA. *J. Low Power Electron. Appl.* **2022**, *12*, 39. [CrossRef]
50. Klass, F.; Weiser, U. Efficient systolic arrays for matrix multiplication. In Proceedings of the International Conference on Parallel Processing, Austin, TX, USA, 12–16 August 1991; Volume III, pp. 21–25.

Disclaimer/Publisher's Note: The statements, opinions and data contained in all publications are solely those of the individual author(s) and contributor(s) and not of MDPI and/or the editor(s). MDPI and/or the editor(s) disclaim responsibility for any injury to people or property resulting from any ideas, methods, instructions or products referred to in the content.

electronics MDPI

Article

FPGA-Flux Proprietary System for Online Detection of Outer Race Faults in Bearings

Jonathan Cureño-Osornio [1], Israel Zamudio-Ramirez [1,2], Luis Morales-Velazquez [1], Arturo Yosimar Jaen-Cuellar [1], Roque Alfredo Osornio-Rios [1] and Jose Alfonso Antonino-Daviu [2,*]

1 CA Mecatrónica, Facultad de Ingeniería, Universidad Autónoma de Querétaro, Av. Río Moctezuma 249, San Juan del Río, Querétaro 76807, Mexico; jcureno08@alumnos.uaq.mx (J.C.-O.); iszara@doctor.upv.es (I.Z.-R.); luis.moralesv@uaq.mx (L.M.-V.); arturo.yosimar.jaen@uaq.mx (A.Y.J.-C.); raosornio@hspdigital.org (R.A.O.-R.)
2 Instituto Tecnológico de la Energía, Universitat Politècnica de València (UPV), Camino de Vera s/n, 46022 Valencia, Spain
* Correspondence: joanda@die.upv.es

Abstract: Online fault detection in industrial machinery, such as induction motors or their components (e.g., bearings), continues to be a priority. Most commercial equipment provides general measurements and not a diagnosis. On the other hand, commonly, research works that focus on fault detection are tested offline or over processors that do not comply with an online diagnosis. In this sense, the present work proposes a system based on a proprietary field programmable gate array (FPGA) platform with several developed intellectual property cores (IPcores) and tools. The FPGA platform together with a stray magnetic flux sensor are used for the online detection of faults in the outer race of bearings in induction motors. The integrated parts comprising the monitoring system are the stray magnetic flux triaxial sensor, several developed IPcores, an embedded processor for data processing, and a user interface where the diagnosis is visualized. The system performs the fault diagnosis through a statistical analysis as follows: First, a triaxial sensor measures the stray magnetic flux in the motor's surroundings (this flux will vary as symptoms of the fault). Second, an embedded processor in an FPGA-based proprietary board drives the developed IPcores in calculating the statistical features. Third, a set of ranges is defined for the statistical features values, and it is used to indicate the condition of the bearing in the motor. Therefore, if the value of a statistical feature belongs to a specific range, the system will return a diagnosis of whether a fault is present and, if so, the severity of the damage in the outer race. The results demonstrate that the values of the root mean square (RMS) and kurtosis, extracted from the stray magnetic field from the motor, provide a reliable diagnostic of the analyzed bearing. The results are provided online and displayed for the user through interfaces developed on the FPGA platform, such as in a liquid crystal display or through serial communication by a Bluetooth module. The platform is based on an FPGA XC6SLX45 Spartan 6 of Xilinx, and the architecture of the modules used are described through hardware description language. This system aims to be an online tool that can help users of induction motors in maintenance tasks and for the early detection of faults related to bearings.

Keywords: embedded systems; intelligent systems; industrial applications; FPGA; reconfigurable computing

Citation: Cureño-Osornio, J.; Zamudio-Ramirez, I.; Morales-Velazquez, L.; Jaen-Cuellar, A.Y.; Osornio-Rios, R.A.; Antonino-Daviu, J.A. FPGA-Flux Proprietary System for Online Detection of Outer Race Faults in Bearings. *Electronics* **2023**, *12*, 1924. https://doi.org/10.3390/electronics12081924

Academic Editor: Alexander Barkalov

Received: 21 March 2023
Revised: 14 April 2023
Accepted: 18 April 2023
Published: 19 April 2023

Copyright: © 2023 by the authors. Licensee MDPI, Basel, Switzerland. This article is an open access article distributed under the terms and conditions of the Creative Commons Attribution (CC BY) license (https://creativecommons.org/licenses/by/4.0/).

1. Introduction

Currently, equipment used for monitoring industrial processes is still of research interest, specifically for the topic of fault detection and classification, since it is continuously being improved thanks to the development of new methodologies for data processing [1]. These systems are very helpful in providing information about a process's status, for instance, indicating whether fault conditions exist, generating alarms, performing error corrections, logging data, and executing critical decisions, among others [2]. Thanks to these systems, corrective actions are avoided, giving way to preventive actions applied in the practice of scheduled maintenance, and, in terms of the benefits, the costs savings are increased and the stopping times of the process are reduced [3]. Most of the monitoring systems are commercial equipment that generally measure common variables, such as the voltage or current, for performing a diagnostic, but they require the knowledge or experience of an expert in the field [4]. Commercial equipment is characterized by being completely closed architectures in hardware and software, and the functionality is also restricted; in fact, the actual tendency in commercial products is to pay for access to the full resources or extra functions [5]. The general scopes of existing monitoring systems are limited because the methodologies implemented for fault detection are insufficient for online implementation [6] or are very complex, requiring many computational resources that need to be implemented offline [7]. Along this same line, the industry has evolved toward a new concept, known as Industry 4.0, that today represents the integration of physical objects, machines, systems, and processes throughout interconnexion networks [8]. The idea is to exploit the full potential of these evolved technologies for performing more than simple measurements through instrumentation techniques. In this sense, to follow the philosophy of Industry 4.0 and achieve smart systems with high profits, monitoring systems are still important elements that must integrate sensors, data processing units, and communication modules for the detection of problems in industrial processes [9]. For these systems, it is very important to have adequate and high-quality information from the processes to accurately detect problems or failures, because some machines, such as an electric motor and its peripheral components, are the most used in industry, since they provide movement and transmit power to the processes, representing between 60% and 80% of the total power consumption [10,11]. In addition, it is worth mentioning that among the elements integrated in an induction motor, the most frequent faults represent between 41 and 42% for rolling bearings, between 28 and 36% for stator winding damages, between 8 and 9% for rotor-related damages, and between 14 and 28% for other types of damages [12,13]. These metrics emphasize that the most common failures in motors appear in the rolling bearings, where several works have focused their efforts on developing methodologies for the detection of problems, considering mainly outer–inner race faults [14,15], ball defects [16], and cage damage [17]. For these reasons, the development of industrial equipment for online fault detection through dedicated methodologies implemented into embedded systems that overcome the limitations and restrictions of existing commercial systems is still an area of opportunity.

Particularly for bearing faults, over time several methodologies have been proposed for detecting such problems. For instance, in the work presented in [18] vibration signature analysis is used together with continuous wavelet transform (CWT) for identifying patterns associated with the vibration signals from bearings in rotating machines. The conditions analyzed were inner and outer race faults, ball faults, and cage faults. The signals were acquired through a very low-cost commercial development platform based on a microcontroller and using a low-cost accelerometer. In addition, the data processing was performed offline with a personal computer (PC) in MATLAB 2013R due to the limited capability of the microcontroller. On the other hand, an acoustic signal analysis was applied offline in [19] for detecting bearing faults in induction motors. In such a proposal, the measured sound of the motor is considered contaminated by other surrounding sources, which have degraded the signal-to-noise ratio (SNR). Therefore, to overcome this situation, a lock-in amplification (LIA) is synchronized to the machine shaft's frequency by means of a fractional phase-locked loop (PLL) frequency synthesizer, yielding the frequency associated to bearing faults. The signals acquisition is performed using a PCIe-6346 National Instruments board and the processing by means of the synthesizer, but the graphical results are presented on the PC. In another case, the use of infrared thermal images demonstrates that fault diagnosis can be performed offline over a rotor-bearing system of a kinematic chain [20]. In that research, several thermal images are acquired from healthy states of the rotor-bearing system and then an exponential linear unit together with stochastic pooling is used to construct an enhanced convolutional neural network (ECNN). In addition, the model parameters of a convolutional auto-encoder (CAE), previously trained with unbalanced images, are transferred to the ECNN; thus, the small labeled thermal images serve to train the ECNN and for the diagnosis of faults. All of the processing was carried out with a PC Core (TM) i7-8550U CPU with 12 GB RAM using the software MATLAB 2016b. The considered conditions for the analysis were normal state, inner and outer race faults, ball faults, and a combination of shaft unbalance with ball faults. On another topic, machine learning and deep learning methodologies have been proposed for fault diagnosis in bearings of rotating machinery. For example, in the research described in [21], raw vibration signals from a kinematic chain are converted into a two-dimensional image in gray scale through their resampling and normalization. In addition, in that work, the use of two dropout layers and two fully connected layers improves the performance of a convolutional neural network (DFCNN) that finally learns the fault patterns, yielding a final diagnostic. The DFCNN methodology was implemented offline on a Ryzen 5 1600X CPU computer with 16 GB RAM and a GTX1060 GPU using MATLAB 2018a software through the neural network toolbox, and the conditions considered were normal state, inner and outer race faults, and ball faults. In other works, such as in the comparative analysis developed in [13], the implementation (on a PC) of deep learning (DL) algorithms to determine which of them were more efficient in detecting bearing faults in mechanical systems was studied and tested. The conclusion from that comparative analysis was that the most popular DL techniques are convolutional neural networks (CNNs) [22], recurrent neural networks (RNNs) [23], auto-encoders (AEs) [24], and generative adversarial networks (GANs) [25]. However, in all previously discussed cases, the implementation of the methodologies was conducted offline using software tools on a PC because of the effort required for the data processing and due to the techniques' complexity, mainly in those data-driven works based on machine learning and deep learning. Currently, there exist some solutions in the field of fault detection in bearings that have been implemented into hardware. For example, the work described in [26] presents an algorithm implemented into a field programmable gate array (FPGA) that performs signature extraction in the time–frequency domain together with a one-against-all multiclass support vector machine for online fault diagnosis in bearings, but a limitation was the computational complexity that restricted the use of the system in real-time applications. Such a methodology uses emitted acoustic signals from a sensor located near the bearing. The analyzed faults were inner and outer race cracks and roller cracks in a cylindrical bearing. In summary, from an analysis of the works reported in the literature, it can be concluded that several methodologies have been developed for fault detection in rolling bearings, but their implementation is limited to offline applications because of the algorithms' complexity. Therefore, to overcome such limitations, FPGAs can be viable alternative solutions for the online processing of algorithms.

Regarding technologies that can be used for developing embedded systems focused on monitoring applications, FPGAs are very advantageous hardware-based development platforms because of their features such as configurability, flexibility, portability, design of modular cores, design of concurrent structures, high speed, high performance, very dedicated design, and hardware description, among others [27]. As an example of the aforementioned, in [28] the power quality issue is addressed through a methodology capable of detecting voltage and current swells by implementing into the FPGA spline interpolation and Otsu segmentation. In such an online implementation, the time-span measurement of the swell disturbance reaches up to 81.3 μs. Through its part, the research described in [29] presents a methodology for measuring the synchronous relationship between electric signals (phase) through a hardware architecture described for an FPGA. This architecture allows to register phase shift changes per minute with a minimum sampling time in the range of picoseconds. This way, the phase measurement core logic unit is based on the subsampling accumulation principle though a systematic sampling over a phase detector. However, this core was validated under a mathematical model. On another topic, the review developed in [30] introduces the evolution and application of different hardware architectures for processing medical imaging through specific technologies. Among the technologies considered in that study of FPGAs are central processing units (CPUs), graphics processing units (GPUs), digital signal processors (DSPs), and application-specific integrated circuits (ASICs), and a discussion of the options according to the application is provided. Recently, a review and survey were presented assessing the implementation of different intelligent techniques, as well as machine learning techniques, for classification tasks in FPGAs [31,32]. In these works, an overview of the wide variety of classification techniques and intelligent techniques is presented, and then the existing FPGA-based implementations of the techniques are discussed; later, the challenges and strategies adopted for the optimization are analyzed, and architectures for hardware accelerators are mentioned. Another survey addressing sensor systems implemented into FPGAs for different applications was developed in [33]. In that work, the assessment was performed for three types of wireless sensor nodes: standalone, combinations of FPGAs with a microcontroller, and FPGA coprocessors for experimental nodes. The objective of the survey was to demonstrate how the features of FPGAs, such as configurability, power consumption, and smart architectures, play a key role in the construction of sensor nodes. An interesting concept that makes use of the potential and advantages provided by FPGAs are called hardware-in-the-loop (HIL) simulations; for instance, in [34] an overview of the engineering advances involving system simulation based on hardware for automotive applications, power electronic systems, and even for industrial drivers is provided. The analysis in that work demonstrates that HIL simulations can reduce the effort during the development and testing of digital systems. On another note, the work described in [35] demonstrates that FPGAs allow for the implementation of reconfigurable architectures in filtering applications under acoustic environments for cancelling noise. To achieve this, the implementation of hardware of flexible finite impulse response (FIR) filters in adaptive linear element (ADALINE) structures are complemented with dedicated multiply accumulate (MAC) units and optimized using least mean square (LMS) and recursive least square (RLS) algorithms. Naturally, the hardware description was optimized, reducing the number of resources in comparison with other implementations. All of these discussed works provide the antecedent that hardware architectures can be implemented into FPGAs for online applications because of their advantages and potential. Therefore, it is desirable to explore the development of an online tool applied for fault diagnosis in bearings based on FPGAs.

The contribution of this work is a methodology for developing a dedicated system based on a field programmable gate array, and it uses stray magnetic flux signals for the online fault detection of the outer race of rolling bearings in induction motors. The monitoring system integrates a proprietary FPGA board and a stray magnetic flux triaxial sensor (which measures the motor's surroundings) for performing data acquisition, processing, and fault detection, making the system nonintrusive. The system performs the fault diagnosis through statistical analysis by measuring the stray magnetic flux in the motor's surroundings (which varies due to the presence of a fault) through a triaxial sensor. Next, the signal is processed through an FPGA-based proprietary board in which an embedded processor drives developed IPcores that calculate the statistical features. After, a set of ranges is defined for the values of the statistical features, and this is used to indicate the condition of the bearing in the motor, considering from a normal condition (without a fault) to the highest severity level. Therefore, if the value of a statistical feature belongs to a specific range, the system will return a diagnosis: whether a fault is present and severity of the damage in the outer race. The developed system can be seen as a digital tool, portable, and nonintrusive for industrial applications that takes advantage of the features of FPGAs for detecting graduality in the faults of bearings. For the validation of the system, it was subjected to tests on damaged bearings whose failures were represented by holes drilled in the outer race; but these holes can represent more than a single fault, for instance, surface breakage and electrical erosion according to the International Organization for Standardization's (ISO) standard. The system developed has the advantages of being configurable, with an open architecture in hardware and software, portable for FPGA technologies and vendors, with a high operating frequency, modularity in the cores' functionalities, and efficiency. The output of the system is adequate from the viewpoint of an industrial product, because it provides to the final users, clearly and concretely, the information of the status detected, for instance, if the bearing is healthy or if it has a fault and its type. However, the information generated by the system can also be used for a deeper analysis, since the acquired signals can be sent to a PC for graphical analysis and other types of data processing, if required. The obtained results demonstrate that the system can provide accurate diagnostics of faults detected in outer races of rolling bearings.

2. Theoretical Foundations

In this section, the theoretical foundations regarding the following topics are described: (i) failures in rolling bearings, (ii) statistical features, and (iii) proprietary boards based on a field programmable gate array.

2.1. Failures in Rolling Bearings

Rolling bearings are important elements used for a wide variety of purposes in industry. However, from all of their possible applications, their implementation in induction motors is the topic of interest of this work, since they are the elements with approximately 40% of the faults in these machines [12,13]. According to [36], bearings are used for transmitting rotating mechanical power in industrial processes through induction machines, and they must accomplish the exacting demands of having a load-carrying capability, running accuracy, noise levels, friction and frictional heat, and life and reliability. Despite the effort in the design and the careful manufacturing of rolling bearings, sometimes their useful lifespan is not fully achieved. As a complement, there exists a standard that explains and classifies the damage and failures occurring in the service of rolling bearings made of standard steels, which is ISO 15243 [37]. From this standard, it is explained that damage and/or failures of these elements can be the result of different circumstances, such as several mechanisms operating simultaneously; improper transport, handling, mounting, and maintenance; faulty manufacturing (of the bearing and adjacent parts); operating conditions; environmental effects; premature failures; aging; cracking; wearing; and corrosion. The consequences are reflected in damage to the elements, economic losses caused by production stoppages, and maintenance and reparation costs. The general classification of the failure modes according to ISO 15243 can be observed in Table 1. It must be specified that the shadowed rows in the table mark the specific failures addressed in this work, which are electrical erosion and fracture/cracking. Outer race failures are common during the operation of rolling bearings, and aging and temperature changes can induce fractures and cracks in the surface. However, current leakage can also cause microspalls in the surface; this phenomenon is also called pitting [36].

Table 1. Damage and failures in rolling bearings according to ISO 15243 [37].

Failure Mode	Failure Subtype
Fatigue	Subsurface initiated fatigue Surface initiated fatigue
Wear	Abrasive wear Adhesive wear
Corrosion	Moisture corrosion Frictional corrosion
Electrical erosion [1]	Excessive current erosion Current leakage erosion
Plastic deformation	Overload deformation Indentations from debris
Fracture and cracking [1]	Forced fracture Fatigue fracture Thermal cracking

[1] Failure modes addressed in this work.

2.2. Statistical Features

A methodology for performing adequate monitoring and detection of faults necessarily requires processing measured signals acquired from the physical system. For such a task, it is very common to carry out the extraction of features from a signal that provide useful information related to the looked for faults. There are many techniques that perform feature extraction; however, in this work, the use of statistical indicators was adopted because they have proven their effectiveness in the development of methodologies for monitoring systems [38,39]. Therefore, in Table 2, a summary of the eleven statistical indicators used in this research is presented. These indicators were obtained directly in the time domain of the measured signals by the monitoring system based in the FPGA. It is worth mentioning that the selection of these features was because they can easily be computed into a hardware structure and provide meaningful information that could be related to faults through patterns, signatures, profiles, and data distribution (central tendencies, dispersion, asymmetries, geometry, and form), which are not always directly visible from the signals.

Table 2. Statistical indicators adopted for this analysis.

Feature	Equation	
Mean	$\overline{x} = \frac{1}{N}\cdot\sum_{i=1}^{N}(x_i)$	(1)
Mean of absolutes	$\overline{x}_a = \frac{1}{N}\cdot\sum_{i=1}^{N}\lvert x_i\rvert$	(2)
Root mean square	$x_{rms} = \sqrt{\frac{1}{N}\cdot\sum_{i=1}^{N}(x_i)^2}$	(3)
Standard deviation	$\sigma = \sqrt{\frac{1}{N}\cdot\sum_{i=1}^{N}(x_i-\overline{x})^2}$	(4)
Variance	$\sigma^2 = \frac{1}{N}\cdot\sum_{i=1}^{N}(x_i-\overline{x})^2$	(5)
RMS shape factor	$SF_{rms} = \frac{x_{rms}}{\overline{x}_a}$	(6)
Maximum value	$x_p = max\lvert x_i\rvert$	(7)
Crest factor	$x_{CF} = \frac{x_p}{x_{rms}}$	(8)
Impulse factor	$x_{IF} = \frac{x_p}{\overline{x}_a}$	(9)
Skewness [1]	$x_{skew} = \frac{\frac{1}{N}\sum_{i=1}^{N}(x_i-\overline{x})^3}{\sigma^3}$	(10)
Kurtosis [1]	$x_{kurt} = \frac{\frac{1}{N}\sum_{i=1}^{N}(x_i-\overline{x})^4}{\sigma^4}$	(11)

[1] High-order moments.

From the table, x is the input data vector from which the statistical features are to be extracted; N is the total number of data in the sample set; and i is the corresponding ith sample, which takes values from $i = 1, 2, 3, \ldots, N$.

2.3. Proprietary Board Based on a Field Programmable Gate Array

The FPGA-based proprietary board in which the monitoring system was implemented has the following features. The proprietary board of 50 × 50 mm dimensions includes one Spartan 6 XC6SLX45 FPGA running at 48 MHz and integrates the power management, static random-access memory (RAM), flash storage, and communication ports, such as the universal serial bus (USB) and universal asynchronous receiver transmitter (UART). The embedded processor, xQuP01v0, and interconnection in-system bus (ISB) in the FPGA are described as follows, both of which are proprietary designs. The processor, xQuP01v0, is a reduced instruction set computer (RISC) structure of a 16-bit core, with its own instruction set architecture (ISA) that integrates a hardware floating-point co-processor for simple precision. The ISB connection is a multiplexed bus protocol for the IPcores interconnection that the embedded processor uses to communicate with the rest of the modules in the system, including the processing modules, communication, and data storage. Both the embedded processor and the ISB are designed to minimize the use of resources in the FPGA, since their objective is to serve as the controller and communication mechanisms of the hardware processing modules. The firmware executed in the embedded processor is used for coordinating the data acquisition, managing the memory, transferring data to the hardware processing modules, and processed data recovery. In addition, the processor is aware of the communications and the user interface. The system uses this hardware–firmware division in order to obtain the maximum performance of the hardware with the software's flexibility, which takes advantage of the versatility of the FPGA to implement fast processing hardware units and complex control processes implemented in software running on the embedded processor. This system has proven to be effective in other applications [40].

3. Proposed Fault Detector Based on FPGA and Stray Flux Applied on Bearings

In this section, the methodology followed for developing a dedicated system for diagnosing faults in rolling bearings (considering outer race faults according to the standard ISO 15243) of induction motors is described. The monitoring tool is implemented into an FPGA-based proprietary board making use of stray magnetic flux signals measured in the motor's surroundings and by computing statistical indicators. Figure 1 presents a block diagram of this tool, named the stray magnetic flux fault detector (SMFFD), and the implementation can be revised in four main blocks: (i) physical system, (ii) stray magnetic flux triaxial sensor, (iii) statistical module implemented in the FPGA-based proprietary board, and (iv) user interface. As can be noted from the figure, the last three blocks integrate the SMFFD system.

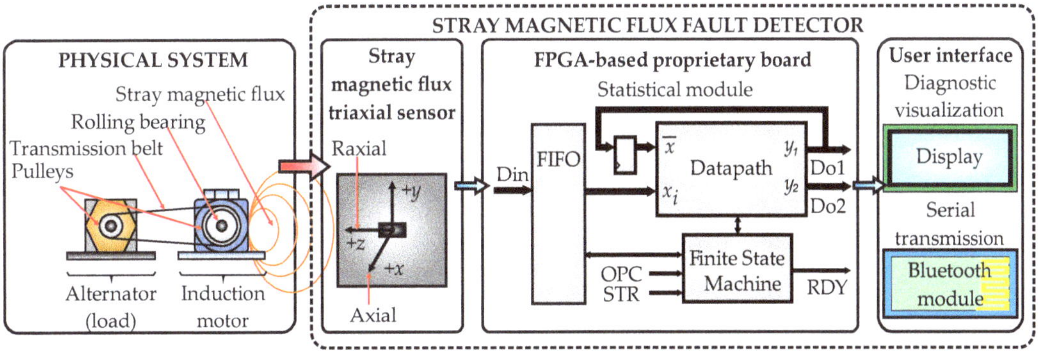

Figure 1. Block diagram of the methodology for detecting faults in the outer race of rolling bearings through the SMFFD.

3.1. Physical System

From Figure 1, the first block of the proposed methodology is the physical system consisting of the electromechanical coupling between an induction motor and an automotive alternator used as the load. The coupling is performed through a transmission belt and two pulleys at the motor and alternator shafts. This way, the system will work under industrial operating conditions. Inside the motor, the rolling bearing under analysis supports the rotor frontal shaft, and during the experimental tests, this bearing is substituted by a bearing with its respective fault condition (healthy and fault conditions). As mentioned in the standard ISO 15243, faults in bearings are associated to different failure modes [37]. However, in this work, only those related with fracture and cracking affecting the outer race of the element were considered because of their statistical importance as reported in the literature. In addition, pitting is an outer race affectation due to the fact of electrical erosion, and this failure mode was also considered. Although these failure modes also consider the inner race affectations on bearings, this work begins by analyzing only outer race faults, with the purpose of keeping a simple structure and its online implementation. Nevertheless, the configurability of the proposed SMFFD could allow to expand its functionality, incorporate other sensors, compute and process additional data, and implement additional algorithms with the aim of detecting other types of faults. Thus, three standard steel bearings, model 6203 2RS, manufactured by SKF, with an external diameter of 40 mm, an internal diameter of 17 mm, and 8 caged balls were prepared for testing the conditions considered, i.e., healthy state and two fault conditions in the outer race with increasing severity. The preparation of the bearings is described as follow: The rubber seals were removed from the back of the bearings and the grease inside was completely cleaned with the help of solvents to remove any grease residue. Next, with the help of metallic clamps, the bearing was secured on the bed of a computerized numerical control (CNC) milling machine for drilling holes on the outer race of each bearing, generating the following fault cases: 3 mm hole (fault severity: 1) and 5 mm hole (fault severity: 2). Once the holes were drilled, an exhaustive cleaning was carried out by blowing compressed air into the bearing's interior; in addition, some solvents were applied to avoid any type of burr that would prevent their correct operation. Finally, BATat-3 grease was applied, which is a high-quality bentone adhesive lubricant designed for the maintenance of bearings operating at high temperatures, and the rubber seals were placed again. The first experimental test used the healthy bearing, and posteriorly the bearing under analysis was changed to one with a fault severity of 1 and next by one with a fault severity of 2. The faults induced defined the controlled experimentations on the physical system, having as advantages the development of rapid tests, adequate and realistic fault design, and variations in the fault severity over short times with the desired graduality. For example, the outer race faults were easily and rapidly induced through a machining process with the desired gradual severity, without the need of waiting for long periods of time over physical system's operation until a real failure occurs. As a counterpart, of course, induced faults cannot reflect the unpredictable ways in which a real fault may occur and affect the physical system. However, the system can be adjusted to be validated under tests of bearings with real faults because of its configurability. It must be clarified that induced faults are real damage to a bearing, and when it is mounted in a physical system, it causes, in consequence, a behavior different from that described with a healthy bearing.

3.2. Stray Magnetic Flux Fault Detector

The second block of the proposed methodology, which is the first part of the SMFFD, is the stray magnetic flux triaxial sensor. Therefore, in order to put the use of the sensor in context, the following must first be mentioned. During the operation of the induction motor, using the healthy bearing, a stray magnetic flux is generated; thus, when this bearing suffers damage, the stray flux has variations in the field magnitude and can be measured. Consequently, a triaxial sensor measures such variations of the stray flux in the motor's surroundings from sensor axes "x", "y", and "z", which for this work corresponded to the axial, axial–radial, and radial directions, respectively. It is worth mentioning that the useful information that the stray magnetic flux can provide will depend on factors such as the element analyzed in the motor, the sensor's placement, and the type of fault studied, among others. However, based on previous experimentations and reported works in the literature [41], axial direction measurement was considered in this work. Finally, the measured signal was sent to the digital system for data processing.

3.3. Statistical Module Implemented into the FPGA-Based Proprietary Board

The third block of the proposed methodology is the statistical module hardware architecture of the SMFFD, which uses the data acquired from the stray magnetic flux sensor to compute the statistical indicators through Equations (1)–(11), as shown in Table 2. To perform this task, the algorithms required in this module are implemented into the FPGA-based proprietary board. Therefore, the hierarchical architecture, developed through hardware description language (HDL), is integrated by the IPcores/soft-cores, such as the first-in, first-out (FIFO) memory, the datapath, a finite states machine (FSM), and pipeline registers. It must be said that the proprietary board implements an embedded processor to control all of the hardware architectures developed for this module and any module required. This embedded processor is not visible in the block diagram in Figure 1, but its function is to drive the data acquisition from the stray magnetic flux sensor, statistical module, and user interface (diagnostic visualization and serial transmission of data). For its part, the FIFO memory stores the input data vector (Din) of the measured signal to speed up the calculation. Meanwhile, the FSM drives the operation of the datapath module and regulates the input of the data from the FIFO. Hence, the embedded processor executes the following sequence, loads the input data to the FIFO memory, defines the operation that the datapath must perform through the OPC command (statistical feature required), and starts the operation process by means of a pulse in the STR terminal. Once the module finishes, it generates a pulse on the RDY terminal and the output data (i.e., result) can be read on terminals Do1 and Do2. In summary, the FSM has the purpose of synchronizing all calculations in case extra steps are required to deliver the expected result.

The main IPcore of the hardware architecture is the datapath, because the statistical features are obtained through this module. Figure 2 shows a simplified diagram of the internal structure of this function, which consists of four main stages: subtraction, powers, accumulator, and adjust. The subtraction stage determines whether the mean, $\overline{x}$, must be subtracted from the input, x_i, or if the absolute value, $|x_i|$, is taken. The powers stage determines the power to which the result of the previous operation is raised, and the power values are 1, 2, 3, and 4. The accumulator stage performs the accumulation of the result of the powers module. For the case of the $\frac{1}{N}$ divisor, this operation is executed as a shift in the fixed-point representation, since N is always considered as an exact power of 2. The adjust stage carries out the rounding and saturation, if applicable, of the accumulated value, and it is the output y_1; in addition, if necessary, it applies the square root of the rounded result, and it is the output y_2. It is important to mention that for every stage, pipeline registers are used with the objective of balancing the latency lines of the computational process.

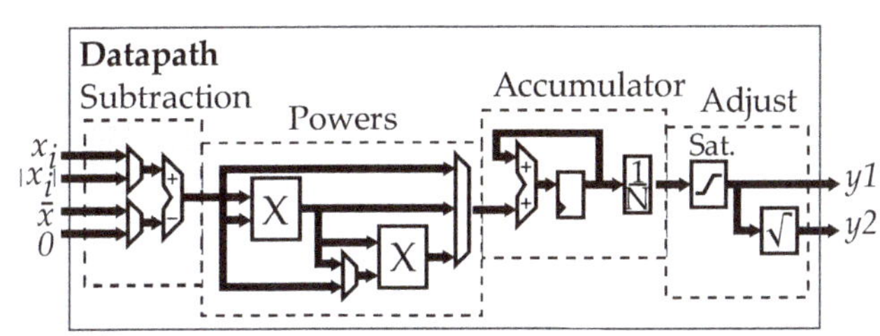

Figure 2. General hardware architecture of the datapath IPcore.

All operations in the datapath IPcore are performed in a fixed-point format, which is specified in the presynthesis during IPcore instantiation, where the numerical representations are adjusted automatically during the elaboration process. In general, this hardware architecture requires two full word multipliers used in the powers stage and an additional one used by the square root unit, which is carried out through a successive approximations register (SAR). In addition, several adders/subtractors are required in the initial stage: accumulation and rounding. Moreover, another register is used for the accumulator and several multiplexers for routing the data flow. The implementation of the datapath considers the latency balance, where each combinational operation is isolated from the next by a pipeline register, which controls the combinational delays within the FPGA's structure and maximizes the operation's frequency. For the multipliers, a latency of 4 clock cycles was considered to maintain data coherence, and balancing registers were placed to synchronize the data paths parallel to the multipliers. Finally, for illustrative purposes and for the sake of not extending too much the explanation of the obtention of the statistical indicators, only two calculus chains are described: root mean square and kurtosis.

Therefore, by taking as the basis the general hardware structure in Figure 2 to calculate the root mean square and using Equation (3) in Table 2, Figure 3 shows the data flow, marked in red, selected by the FSM through the OPC command to obtain this value. From the figure, the calculus chain starts by subtracting the mean, $\overline{x}$, to the input, x_i, and this value is then raised to the power of 2, the result is accumulated and divided by N, the square root is obtained from the rounded value, and the output is in terminal y_2. In this case, the powers stage considers a latency of 8 cycles to maintain datapath synchronization.

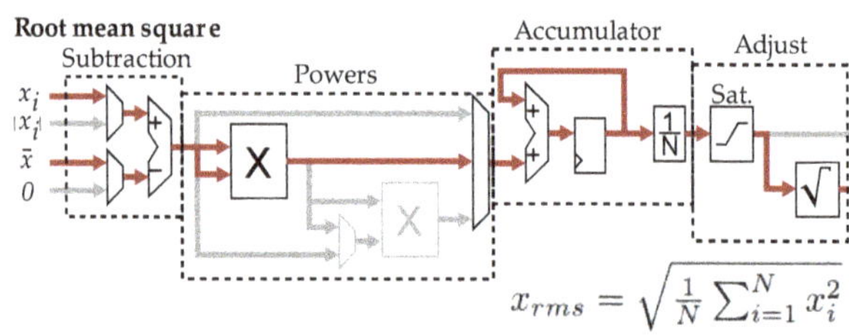

Figure 3. Calculus chain to obtain the root mean square of Equation (3).

Similarly, using the previous example and considering the general diagram in Figure 2 and Equation (11) in the Table 2 to calculate the kurtosis, four steps are necessary: calculation of the mean, calculation of the standard deviation, calculation of the numerator of the kurtosis, and calculation of the kurtosis. Figure 4 shows the data flow, marked in red, to calculate the kurtosis numerator. For this purpose, two multipliers are used in the powers stage to obtain the fourth power term. The final kurtosis value is calculated through the firmware of the floating-point co-processor, in the hardware, by dividing the kurtosis numerator by the standard deviation raised to a power of 4.

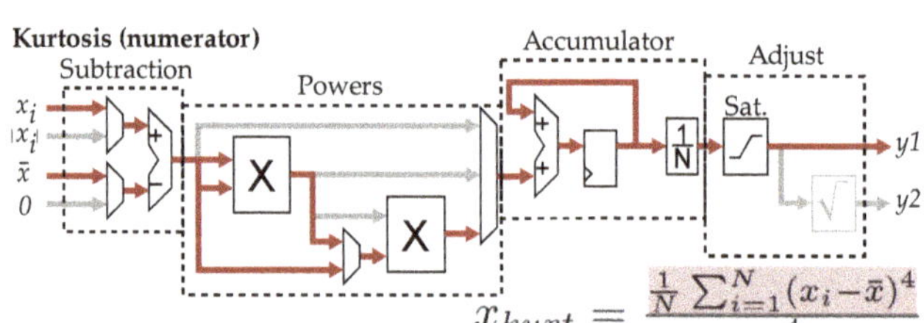

Figure 4. Calculus chain to obtain the kurtosis numerator of Equation (11).

3.4. User Interface

The last block of the proposed methodology is the user interface that presents a visualization of the diagnosis results through a liquid crystal display (LCD). In this LCD screen are displayed the statistical indicators obtained by the SMFFD system. Only those indicators that can provide useful information about the faults detected are displayed on the screen. It is worth mentioning that the user can extract the measured signal and the statistical indicators through an additional port of serial communication, since the system has IPcores to drive a Bluetooth module.

4. Results and Discussion

4.1. Experimental Setup

The experimental test bench is an electromechanical system consisting of the coupling between an induction motor and an automotive alternator. Figure 5a presents a picture of the experimental test bench. The characteristics of the induction machine are as follows: manufactured by WEG, triphasic motor, one pair of poles, case type A.E. 00136AP3E48TCT, rated power of 740 W, nominal speed of 3355 RPM, input voltage of 210-230/460 Vac, and operating frequency of 50/60 Hz. For its part, an automotive alternator was used as the mechanical load entailing approximately 30% of the motor capacity. All elements tested were standard steel bearings manufactured by SKF, model 6203 2RS, with an external diameter of 40 mm, internal diameter of 17 mm, and eight caged balls. The bearings' preparation was described previously in Section 2. Figure 5b presents pictures of the bearings used after such preparation in the three conditions analyzed: healthy bearing and two bearings with severity levels 1 (3 mm hole) and 2 (5 mm hole). As mentioned, inside the motor the bearing under analysis supports the rotor frontal shaft, and during the experimental tests, this bearing is substituted by the bearings with their respective defined condition. According to the standard ISO 15243, these induced faults can be categorized as fractures and cracks but also as pitting phenomenon caused by electrical erosion. For the experimental trials, the electromechanical system was driven through a variable frequency driver (VFD) feeding the motor with a start ramp of 10 s, which was previously programmed, that reached a final operating frequency of 50 Hz. Taking into consideration this ramp, every trial lasted 40 s, with the first 10 s corresponding to the transient response and the last 30 s to the steady state. For the data processing in the SMFFD, only the steady state was considered. This way, a total of 15 runs per bearing condition (healthy state and two outer race severities) were carried out, generating 15 runs $\times$ 3 conditions = 45 data sets.

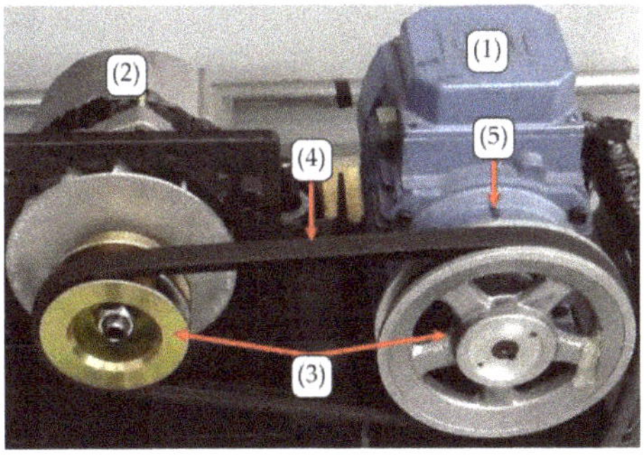

(a)

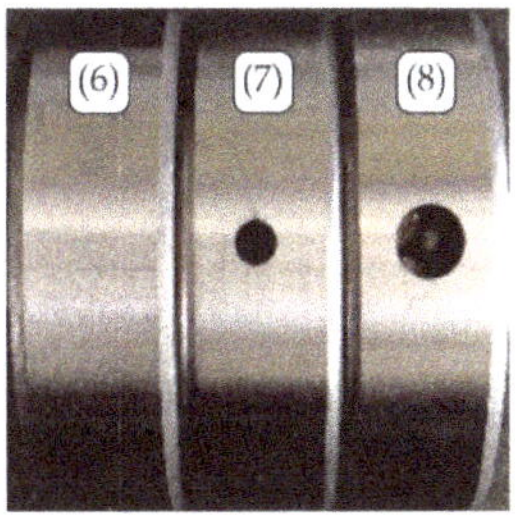

(b)

Figure 5. Experimental setup: (**a**) test bench; (**b**) rolling bearings conditions. The considered elements were the (1) induction motor, (2) alternator as the motor load, (3) output shaft pulleys, (4) transmission belt, (5) location of the bearing under analysis in the frontal support of the shaft, (6) bearing with a healthy outer race, (7) bearing with a 3 mm hole in the outer race, and (8) bearing with a 5 mm hole in the outer race.

4.2. Stray Magnetic Flux Fault Detector

Regarding this tool developed, Figure 6a presents the physical SMFFD system for diagnosing bearings in induction motors. From the figure, it can be noted that in the final package's presentation, only the LCD screen is visible to the user. A sticker indicates the way of placing the SMFFD in respect to the motor for a correct analysis; as previously mentioned, the axial flux ("x"-axis) was of interest [41]. Internally, as shown in Figure 6b, the box contains the FPGA-based proprietary board, Bluetooth module, power source, liquid crystal display, and triaxial stray magnetic flux sensor. For its part, the triaxial sensor for measuring the stray magnetic flux is the board BM1422AGMV-EVK-001 from ROHM Semiconductor manufacturer, and it was installed in the SMFFD box making the "*x*"-, "*y*"-, and "*z*"-axes coincident with the axial, axial–radial, and radial directions of the stray flux generated by the motor, respectively. These sensors had the following features: bandwidth of 1 kHz, I2C interface, sensitivity of 0.042 μT/LSB, sensing range of ±1200 μT, and supply voltage of 1.7–3.6 V. The proprietary board characteristics were described previously in Section 3. In relation to the FPGA, the proposed IPcores were implemented as hardware processing units: inter-integrated circuit (I2C) communication for acquiring the data of the stray magnetic flux sensor, statistical module, LCD driver, communication port UART, and embedded system that comprises the processor, memory driver, ISB connection, and USB interface of the programming. Here, the firmware controls all modules for implementing fault detection through the stray magnetic flux.

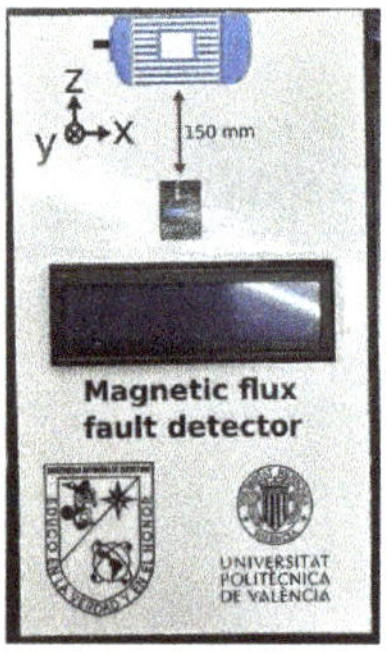

(**a**)

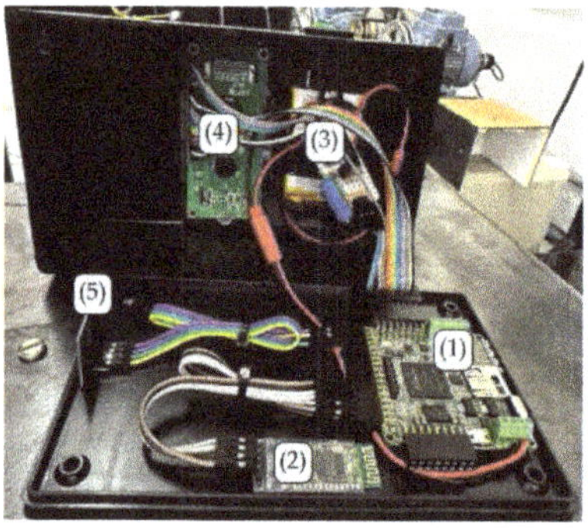

(**b**)

Figure 6. FPGA-based proprietary SMFD: (**a**) physical system; (**b**) hardware components. The hardware components of the SMFD are the (1) FPGA-based proprietary board, (2) Bluetooth module, (3) power source, (4) liquid crystal display, and (5) proprietary triaxial stray magnetic flux sensor.

The SMFFD performed the data acquisition from the sensor at a sampling frequency of 1 kHz, and the data of interest were in the steady state of the machine, as previously mentioned, in the last 30 s of each trial. For the computation of the statistical indicators time windows of 4086 data points were taken with overlaps between the windows of 50%. Therefore, every statistical indicator was obtained and updated approximately every two seconds during the online monitoring process. In this way, 29 indicators were generated per trial, 345 indicators per bearing condition, and a total of 1305 indicators for all three conditions, which were used for validating the diagnosis. The SMFFD monitoring tool indicates in the LCD the information regarding the final diagnosis through the values of the statistical indicators; for instance, on the screen two indicators per row are displayed. The fault severity is known according to a defined range of values to which the statistical indicators belong. Additionally, the measured signal and the statistical indicators can be extracted by the user through serial transmission of the Bluetooth module.

In summary, the Spartan 6 XC6SLX45 is a cost-optimized FPGA, according to the manufacturer, and the hardware resources used by the SMFFD system are presented in Table 3. The resources used consider all of the modules described previously (embedded processor, statistical module, drives for LCD and Bluetooth, etc.). From the table, the column "Logic utilization" refers to the specific hardware elements in the FPGA; the column "Used" indicates the exact number of implemented elements; the column "Available" indicates the total available of each type of element; and the column "Utilization" represents the percentage of elements used in respect to the total available. Finally, the tool for synthetizing the project was Xilinx ISE 14.7, using the Ubuntu 22.04 operating system.

Table 3. Hardware resources of the Spartan 6 XC6SLX45 FPGA used by the SMFFD system.

Logic Utilization	Used	Available	Utilization
Number of slice registers	5401	54,576	9%
Number of slice look up tables (LUTs)	7367	27,288	26%
Number of bonded input–output blocks (IOBs)	103	218	47%
Number of block random access memory (RAM)/the first-in, first-out (FIFO)	17	116	14%
Number of digital signal processing multipliers (DSP48A1s)	14	58	24%

In the case of IOBs, this refers to the physical terminals of the device, where most are IOs connected to external RAM (40 pins), and there are general purpose IOs (16 pins), the rest are connected to the LCD, sensor, communication ports, etc. In general, it can be said that a quarter of the device's resources are used.

4.3. Results of the Fault Diagnosis through the SMFFD

In the next paragraphs, the fault diagnosis through the SMFFD is described. For performing the diagnosis task, thresholds must be defined as follows: The statistical features' values vary in a range according to the bearing's condition in the motor; for instance, the motor with a bearing fault and with a severity level will cause variations in the statistical values different from those when the motor bearing is healthy. Therefore, after several experimental trials, it was found that from the eleven statistical indicators only a few of them present meaningful information related to the fault and its severity; the rest have incipient variations in their values. For this work, the statistical indicators of the RMS and kurtosis provided the best information related to the fault and its severity. Thus, the RMS was used to provide a threshold of whether the kurtosis data were valid, because if the RMS was out of range, then the kurtosis was saturated. This range, which was experimentally defined, established the RMS amplitude between 150 and 250, which corresponds to measurements of the sensor between 5 μT and 10 μT, respectively. Thus, if the kurtosis values fell into this range, then data were valid, but for values outside of this range, the kurtosis would not be valid. Meanwhile, the kurtosis was used for defining a set of ranges for indicating the bearing's condition. To obtain these kurtosis ranges, an independently short experimentation was carried out as follows: The stray magnetic flux signal was acquired in the time domain (1 kHz sampling frequency) at the steady state and a total of 140 time windows were used, each one of 4 s in length, for every bearing condition (healthy, 3 mm hole, and 5 mm hole). Therefore, 140 windows per three conditions resulted in 420 windows that were used to obtain the kurtosis boxplots for differentiating every bearing condition, see Figure 7.

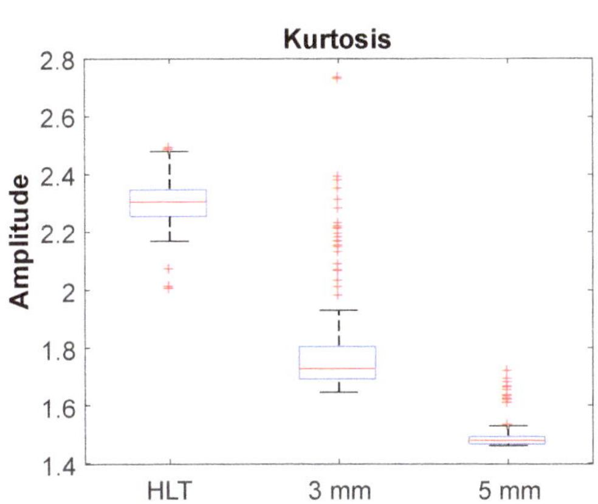

Figure 7. Boxplots that determine the range of kurtosis for every condition of the bearing.

Through the boxplots in Figure 7, it can be noted that on the "*x*"-axis the labels "HLT", "3 mm", and "5 mm" represent the bearing conditions healthy, severity 1, and severity 2, respectively. On the "*y*"-axis are the amplitude values that every boxplot spans. Congruently, the ranges that determine the severity of the faults are summarized in Table 4.

Table 4. Ranges for condition detection according to the kurtosis statistical indicator.

Bearing Condition	Kurtosis Range
Healthy state (HLT)	[2.2–2.5]
Fault severity 1—Hole of 3 mm diameter	[1.5–1.9]
Fault severity 2—Hole of 5 mm diameter	[1.3–1.5]

For the sake of validating the monitoring system's functionality, the results of the three analyzed conditions are presented. Figure 8a shows an image of the SMFFD system performing online monitoring and diagnosis of the induction motor with a healthy state bearing (element without problems in the outer race) mounted in the rotor's frontal shaft. From the picture, it can be observed that the placement of the SMFFD in relation to the induction motor was nonintrusive, because the SMFFD stayed in the surroundings of the physical system. Meanwhile, Figure 8b presents a digital zoom in on the SMFFD focused on the LCD screen to better appreciate the results obtained from the fault diagnosis. From this zoom, the statistical indicators shown to the user are the RMS and kurtosis, having magnitudes of 155.595 (kurtosis is a valid value) and 2.5, respectively. Therefore, by taking the value of the kurtosis and comparing it with the ranges in Table 4, the diagnosis is a healthy bearing. The letter "A" that appears on the LCD screen next to the RMS value is not an indication of units, instead this letter indicates that the SMFFD system is performing the analysis. Additionally, the letter "V" appears next to the value of the kurtosis, and this letter indicates that a fault condition was detected. It is worth highlighting, again, the practicality of the monitoring tool because it has the following advantages: it is a nonintrusive system (the sensor in the SMFFD measures the stray magnetic flux in the motor's surroundings), the fault diagnosis is completely online, the system is portable, its reconfigurability, and the functionality expansion.

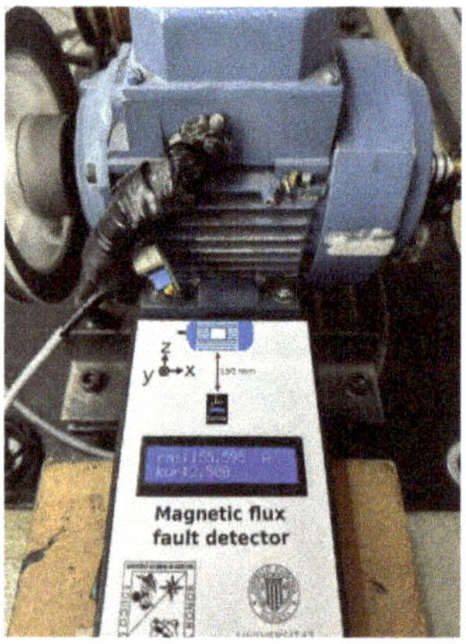

(a)

(b)

Figure 8. (**a**) SMFFD performing an online fault diagnosis on the induction motor with the healthy bearing; (**b**) observing the results in detail through the digital zoom on the LCD screen.

In another case, Figure 9a presents a captured image of the SMFFD system performing online monitoring and diagnosis of the induction motor with a bearing, mounted in the rotor's frontal shaft, having the severity of 1 for an outer race fault (hole of 3 mm). At the same time, Figure 9b presents a digital zoom of the LCD screen for this case, indicating an RMS with a value of 233.105 (kurtosis is a valid value) and the kurtosis with a value of 1.545. Hence, by taking the value of the kurtosis and comparing it with the ranges in Table 4, the diagnosis is effectively the outer race fault on the bearing with a hole of 3 mm.

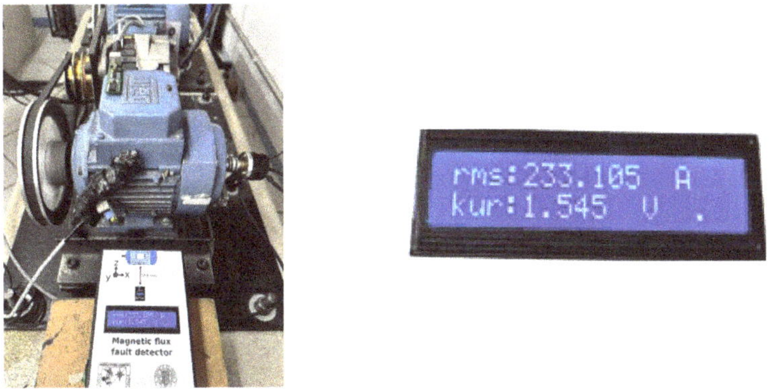

(**a**) (**b**)

Figure 9. (**a**) SMFFD performing an online fault diagnosis on the induction motor with the bearing having a fault severity of 1 (3 mm); (**b**) observing the results in detail through the digital zoom of the LCD screen.

Figure 10a depicts a photograph of the SMFFD system performing online monitoring and diagnosis of the induction motor with a bearing, mounted in the rotor's frontal shaft, having the severity of 2 for an outer race fault (hole of 5 mm). Figure 10b shows a digital zoom on the screen of the tool indicating magnitudes for the RMS of 163.859 (kurtosis is a valid value) and a magnitude for the kurtosis of 1.333. Therefore, by taking the value of the kurtosis and comparing it with the ranges in Table 4, the diagnosis is an outer race fault on the bearing with a hole of 5 mm.

(**a**) (**b**)

Figure 10. (**a**) SMFFD performing an online fault diagnosis on the induction motor with the bearing having a fault severity of 2 (5 mm); (**b**) observing the results in detail through the digital zoom of the LCD screen.

5. Conclusions

This work presents a methodology for developing an online monitoring tool for diagnosing outer race faults on bearings in induction motors. The monitoring tool, named the stray magnetic flux fault detector, was developed into an FPGA-based proprietary board, performing an analysis on the data from the acquired stray magnetic flux signal. The are several advantages to using an FPGA-based solution, for example, the design in the hardware for faster data processing, concurrent execution of the IPcores, configurability, portability, functionality expansion according to the application requirements, and high operational frequency, among others. The practicality of the developed tool is observed in its compact design following an all-in-one philosophy, which means that the system includes the FPGA, an embedded processor, the sensor, the user interface, and the power source. Therefore, the user must only put the system near to the induction motor that needs to be analyzed, and the online fault diagnosis will be performed. This is achieved thanks to the sensor integrated into the system's box, thus measuring the stray magnetic flux from three directions (axial, radial, and axial–radial) in the motor's surroundings, making the system nonintrusive. Now, in relation to the FPGA potential, the implementation of the IPcore for the calculation of the statistical indicators demonstrates the powerfulness of the programmable logic device, because the system acquires the physical signal and extracts the features related to the faults. As mentioned, the selection of the statistical features is because they are relatively easy to compute and can provide nonvisible information about the data distribution. In this sense, the calculation of the statistical indicators was performed through a generalized hardware architecture in only four stages. It is worth mentioning the potential of computing diverse statistical indicators, for this work from the set of statistical features, two of which became meaningful in the final diagnosis. The RMS validates the kurtosis value and this allows for the differentiation of the bearing conditions; thus, the system is based on these two features. However, if other types of faults need to be analyzed, the rest of the statistical features could be helpful, since the faults could be reflected as different symptoms in the physical system and, consequently, in the acquired data. Thus, the calculation of several statistical indicators is important, because they could be useful for the analysis of other types of faults. In addition, in future work other nonstatistical indicators can be computed and explored for developing fault detection methodologies. Finally, as mentioned, the developed system has an interesting and important characteristic which is the configurability allowing for the inclusion of extra IPcores, allowing the system to be adjusted as required. For this reason, the system is able to be expanded in functionality and with the possibility of being explored for other monitoring applications, because other types of sensors can be added and other hardware structures can be described according to an application's requirements.

Author Contributions: Conceptualization, R.A.O.-R. and J.A.A.-D.; methodology, L.M.-V.; software, L.M.-V.; validation, J.C.-O. and I.Z.-R.; formal analysis, J.C.-O.; investigation, A.Y.J.-C.; data curation, J.C.-O. and I.Z.-R.; writing—original draft preparation, A.Y.J.-C. and L.M.-V.; writing—review and editing, A.Y.J.-C.; visualization, R.A.O.-R.; supervision, R.A.O.-R. and J.A.A.-D.; project administration, J.A.A.-D.; funding acquisition, J.A.A.-D. All authors have read and agreed to the published version of the manuscript.

Funding: This research was funded by the Spanish "Ministerio de Ciencia e Innovación", Agencia Estatal de Investigación and FEDER program in the framework of the "Proyectos de Generación de Conocimiento 2021" of the "Programa Estatal para Impulsar la Investigación Científico-Técnica y su Transferencia", belonging to the "Plan Estatal de Investigación Científica, Técnica y de Innovación 2021-2023" (ref: PID2021-122343OB-I00).

Data Availability Statement: Not applicable.

Acknowledgments: CONACyT scholarship.

Conflicts of Interest: The authors declare no conflict of interest.

References

1. Melo, A.; Câmara, M.M.; Clavijo, N.; Pinto, J.C. Open Benchmarks for Assessment of Process Monitoring and Fault Diagnosis Techniques: A Review and Critical Analysis. *Comput. Chem. Eng.* **2022**, *165*, 107964. [CrossRef]
2. Tidriri, K.; Chatti, N.; Verron, S.; Tiplica, T. Bridging Data-Driven and Model-Based Approaches for Process Fault Diagnosis and Health Monitoring: A Review of Researches and Future Challenges. *Annu. Rev. Control* **2016**, *42*, 63–81. [CrossRef]
3. Geng, S.; Wang, X. Predictive Maintenance Scheduling for Multiple Power Equipment Based on Data-Driven Fault Prediction. *Comput. Ind. Eng.* **2022**, *164*, 107898. [CrossRef]
4. Entezami, M.; Hillmansen, S.; Weston, P.; Papaelias, M.P. Fault Detection and Diagnosis within a Wind Turbine Mechanical Braking System Using Condition Monitoring. *Renew. Energy* **2012**, *47*, 175–182. [CrossRef]
5. Cheshmeh, Z.A.; Bigverdi, Z.; Eqbalpour, M.; Kowsari, E.; Ramakrishna, S.; Gheibi, M. A Comprehensive Review of Used Electrical and Electronic Equipment Management with a Focus on the Circular Economy-Based Policy-Making. *J. Clean. Prod.* **2023**, *389*, 136132. [CrossRef]
6. Dong, L.; Liu, S.; Zhang, H. A Method of Anomaly Detection and Fault Diagnosis with Online Adaptive Learning under Small Training Samples. *Pattern Recognit.* **2017**, *64*, 374–385. [CrossRef]
7. Seera, M.; Lim, C.P.; Ishak, D.; Singh, H. Offline and Online Fault Detection and Diagnosis of Induction Motors Using a Hybrid Soft Computing Model. *Appl. Soft Comput.* **2013**, *13*, 4493–4507. [CrossRef]
8. Hajoary, P.K. Industry 4.0 Maturity and Readiness- A Case of a Steel Manufacturing Organization. *Procedia Comput. Sci.* **2023**, *217*, 614–619. [CrossRef]
9. Mykoniatis, K. A Real-Time Condition Monitoring and Maintenance Management System for Low Voltage Industrial Motors Using Internet-of-Things. *Procedia Manuf.* **2020**, *42*, 450–456. [CrossRef]
10. Ghosh, P.K.; Sadhu, P.K.; Basak, R.; Sanyal, A. Energy Efficient Design of Three Phase Induction Motor by Water Cycle Algorithm. *Ain Shams Eng. J.* **2020**, *11*, 1139–1147. [CrossRef]
11. Boteler, R.; Malinowski, J. Review of Upcoming Changes to Global Motor Efficiency Regulations. In Proceedings of the Conference Record of 2009 Annual Pulp and Paper Industry Technical Conference, Birmingham, AL, USA, 21–26 June 2009; pp. 26–30.
12. Gangsar, P.; Tiwari, R. Signal Based Condition Monitoring Techniques for Fault Detection and Diagnosis of Induction Motors: A State-of-the-Art Review. *Mech. Syst. Signal Process.* **2020**, *144*, 106908. [CrossRef]
13. Hakim, M.; Omran, A.A.B.; Ahmed, A.N.; Al-Waily, M.; Abdellatif, A. A Systematic Review of Rolling Bearing Fault Diagnoses Based on Deep Learning and Transfer Learning: Taxonomy, Overview, Application, Open Challenges, Weaknesses and Recommendations. *Ain Shams Eng. J.* **2023**, *14*, 101945. [CrossRef]
14. Kumar, P.; Kumar, P.; Hati, A.S.; Kim, H.S. Deep Transfer Learning Framework for Bearing Fault Detection in Motors. *Mathematics* **2022**, *10*, 4683. [CrossRef]
15. Gao, Y.; Yu, D. Fault Diagnosis of Rolling Bearing Based on Laplacian Regularization. *Appl. Soft Comput.* **2021**, *111*, 107651. [CrossRef]
16. Jayakanth, J.J.; Chandrasekaran, M.; Pugazhenthi, R. Impulse Excitation Analysis of Material Defects in Ball Bearing. *Mater. Today Proc.* **2021**, *39*, 717–724. [CrossRef]
17. Wen, C.; Meng, X.; Fang, C.; Gu, J.; Xiao, L.; Jiang, S. Dynamic Behaviors of Angular Contact Ball Bearing with a Localized Surface Defect Considering the Influence of Cage and Oil Lubrication. *Mech. Mach. Theory* **2021**, *162*, 104352. [CrossRef]
18. Saxena, M.; Bannet, O.O.; Gupta, M.; Rajoria, R.P. Bearing Fault Monitoring Using CWT Based Vibration Signature. *Procedia Eng.* **2016**, *144*, 234–241. [CrossRef]
19. Del Rosario Bautista-Morales, M.; Patiño-López, L.D. Acoustic Detection of Bearing Faults through Fractional Harmonics Lock-in Amplification. *Mech. Syst. Signal Process.* **2023**, *185*, 109740. [CrossRef]
20. Zhiyi, H.; Haidong, S.; Xiang, Z.; Yu, Y.; Junsheng, C. An Intelligent Fault Diagnosis Method for Rotor-Bearing System Using Small Labeled Infrared Thermal Images and Enhanced CNN Transferred from CAE. *Adv. Eng. Inform.* **2020**, *46*, 101150. [CrossRef]
21. Zhang, J.; Sun, Y.; Guo, L.; Gao, H.; Hong, X.; Song, H. A New Bearing Fault Diagnosis Method Based on Modified Convolutional Neural Networks. *Chin. J. Aeronaut.* **2020**, *33*, 439–447. [CrossRef]
22. Ruan, D.; Wang, J.; Yan, J.; Gühmann, C. CNN Parameter Design Based on Fault Signal Analysis and Its Application in Bearing Fault Diagnosis. *Adv. Eng. Inform.* **2023**, *55*, 101877. [CrossRef]
23. An, Z.; Li, S.; Wang, J.; Jiang, X. A Novel Bearing Intelligent Fault Diagnosis Framework under Time-Varying Working Conditions Using Recurrent Neural Network. *ISA Trans.* **2020**, *100*, 155–170. [CrossRef] [PubMed]
24. Luo, S.; Huang, X.; Wang, Y.; Luo, R.; Zhou, Q. Transfer Learning Based on Improved Stacked Autoencoder for Bearing Fault Diagnosis. *Knowl.-Based Syst.* **2022**, *256*, 109846. [CrossRef]
25. Liu, S.; Jiang, H.; Wu, Z.; Li, X. Rolling Bearing Fault Diagnosis Using Variational Autoencoding Generative Adversarial Networks with Deep Regret Analysis. *Measurement* **2021**, *168*, 108371. [CrossRef]
26. Kang, M.; Kim, J.; Kim, J.-M. An FPGA-Based Multicore System for Real-Time Bearing Fault Diagnosis Using Ultrasampling Rate AE Signals. *IEEE Trans. Ind. Electron.* **2015**, *62*, 2319–2329. [CrossRef]
27. Magyari, A.; Chen, Y. Review of State-of-the-Art FPGA Applications in IoT Networks. *Sensors* **2022**, *22*, 7496. [CrossRef]
28. Lopez-Ramirez, M.; Ledesma-Carrillo, L.M.; Rodriguez-Donate, C.; Miranda-Vidales, H.; Mata-Chavez, R.I.; Cabal-Yepez, E. FPGA-Based Online Voltage/Current Swell Segmentation and Measurement. *Comput. Electr. Eng.* **2023**, *107*, 108620. [CrossRef]

29. Mitra, J.; Nayak, T.K. An FPGA-Based Phase Measurement System. *IEEE Trans. Very Large Scale Integr. VLSI Syst.* **2018**, *26*, 133–142. [CrossRef]
30. Alcaín, E.; Fernández, P.R.; Nieto, R.; Montemayor, A.S.; Vilas, J.; Galiana-Bordera, A.; Martinez-Girones, P.M.; Prieto-de-la-Lastra, C.; Rodriguez-Vila, B.; Bonet, M.; et al. Hardware Architectures for Real-Time Medical Imaging. *Electronics* **2021**, *10*, 3118. [CrossRef]
31. Saidi, A.; Ben Othman, S.; Dhouibi, M.; Ben Saoud, S. FPGA-Based Implementation of Classification Techniques: A Survey. *Integration* **2021**, *81*, 280–299. [CrossRef]
32. Seng, K.P.; Lee, P.J.; Ang, L.M. Embedded Intelligence on FPGA: Survey, Applications and Challenges. *Electronics* **2021**, *10*, 895. [CrossRef]
33. De la Piedra, A.; Braeken, A.; Touhafi, A. Sensor Systems Based on FPGAs and Their Applications: A Survey. *Sensors* **2012**, *12*, 12235–12264. [CrossRef]
34. Mihalič, F.; Truntič, M.; Hren, A. Hardware-in-the-Loop Simulations: A Historical Overview of Engineering Challenges. *Electronics* **2022**, *11*, 2462. [CrossRef]
35. Ezilarasan, M.R.; Britto Pari, J.; Leung, M.-F. Reconfigurable Architecture for Noise Cancellation in Acoustic Environment Using Single Multiply Accumulate Adaline Filter. *Electronics* **2023**, *12*, 810. [CrossRef]
36. Bearing Failure and How to Prevent It | SKF. Available online: https://www.skf.com/us/products/rolling-bearings/bearing-failure-and-how-to-prevent-it (accessed on 13 March 2023).
37. *ISO 15243:2017(En)*; Rolling Bearings—Damage and Failures—Terms, Characteristics and Causes. ISO 2017, Technical Committee ISO/TC 4 Rolling bearings. Available online: https://www.iso.org/obp/ui/#iso:std:iso:15243:ed-2:v1:en (accessed on 13 March 2023).
38. Alfredo Osornio-Rios, R.; Yosimar Jaen-Cuellar, A.; Ivan Alvarado-Hernandez, A.; Zamudio-Ramirez, I.; Armando Cruz-Albarran, I.; Alfonso Antonino-Daviu, J. Fault Detection and Classification in Kinematic Chains by Means of PCA Extraction-Reduction of Features from Thermographic Images. *Measurement* **2022**, *197*, 111340. [CrossRef]
39. Jaen-Cuellar, A.Y.; Trejo-Hernández, M.; Osornio-Rios, R.A.; Antonino-Daviu, J.A. Gear Wear Detection Based on Statistic Features and Heuristic Scheme by Using Data Fusion of Current and Vibration Signals. *Energies* **2023**, *16*, 948. [CrossRef]
40. Clemente-Lopez, D.; Rangel-Magdaleno, J.J.; Munoz-Pacheco, J.M.; Morales-Velazquez, L. A Comparison of Embedded and Non-Embedded FPGA Implementations for Fractional Chaos-Based Random Number Generators. *J. Ambient Intell. Hum. Comput.* **2022**. [CrossRef]
41. Vitek, O.; Janda, M.; Hajek, V.; Bauer, P. Detection of Eccentricity and Bearings Fault Using Stray Flux Monitoring. In Proceedings of the 8th IEEE Symposium on Diagnostics for Electrical Machines, Power Electronics & Drives, Bologna, Italy, 5–8 September 2011; pp. 456–461.

Disclaimer/Publisher's Note: The statements, opinions and data contained in all publications are solely those of the individual author(s) and contributor(s) and not of MDPI and/or the editor(s). MDPI and/or the editor(s) disclaim responsibility for any injury to people or property resulting from any ideas, methods, instructions or products referred to in the content.

electronics

MDPI

Article

Finding the Top-K Heavy Hitters in Data Streams: A Reconfigurable Accelerator Based on an FPGA-Optimized Algorithm

Ali Ebrahim

Department of Computer Engineering, University of Bahrain, Sakhir P.O. Box 32038, Bahrain; ahasan@uob.edu.bh; Tel.: +973-17437029

Abstract: This paper presents a novel approach for accelerating the top-k heavy hitters query in data streams using Field Programmable Gate Arrays (FPGAs). Current hardware acceleration approaches rely on the direct and strict mapping of software algorithms into hardware, limiting their performance and practicality due to the lack of hardware optimizations at an algorithmic level. The presented approach optimizes a well-known software algorithm by carefully relaxing some of its requirements to allow for the design of a practical and scalable hardware accelerator that outperforms current state-of-the-art accelerators while maintaining near-perfect accuracy. This paper details the design and implementation of an optimized FPGA accelerator specifically tailored for computing the top-k heavy hitters query in data streams. The presented accelerator is entirely specified at the C language level and is easily reproducible with High-Level Synthesis (HLS) tools. Implementation on Intel Arria 10 and Stratix 10 FPGAs using Intel HLS compiler showed promising results—outperforming prior state-of-the-art accelerators in terms of throughput and features.

Keywords: top-k heavy hitters; data streams; Field Programmable Gate Arrays; High-Level Synthesis

1. Introduction

Extracting a list of the most frequently occurring items (aka. heavy hitters) from large datasets is a well-studied problem that is usually tackled with approximation techniques due to the complexity and size of the problem [1,2]. Several approximation techniques model the input as a "data stream" consisting of a sequence of items that needs to be processed in a one-pass manner at high speed and using limited memory [3]. The heavy hitter problem has applications in many fields, such as network traffic monitoring [4], website data analysis [5], and sensor networks [6]. A sub-class of the heavy hitter problem is the "top-k" problem, wherein a user would query the k most frequent items in a data stream. Examples of such queries include the top-visited websites in web data, the most frequent destination IPs in network traffic passing through a networking device, the bestselling products in retail data, etc.

In recent years, data stream algorithms have been deployed by companies such as Google, Apple, Microsoft, etc. to address several computational problems [7]. With the growing demand for high-speed data stream processing, several custom hardware accelerator architectures have emerged (see Section 2). In general, these accelerators rely on parallelism and deep pipelining to achieve the required processing throughputs. Field Programmable Gate Arrays (FPGAs) are typically used to implement such accelerators due to three main reasons: First, the flexibility of FPGAs allows one to change the accelerator hardware configuration so that it is tailored for specific stream distributions or specific user requirements. For example, some hardware configurations would favor accuracy, while other configurations would favor higher throughputs. Second, stream algorithms usually summarize the properties of the data stream in small data structures referred to as "stream summaries". The amount of fast on-chip SRAM memory in a modern FPGA is sufficient for

Citation: Ebrahim, A. Finding the Top-K Heavy Hitters in Data Streams: A Reconfigurable Accelerator Based on an FPGA-Optimized Algorithm. *Electronics* **2023**, *12*, 2376. https://doi.org/10.3390/electronics12112376

Academic Editor: Raffaele Giordano

Received: 11 April 2023
Revised: 13 May 2023
Accepted: 23 May 2023
Published: 24 May 2023

Copyright: © 2023 by the author. Licensee MDPI, Basel, Switzerland. This article is an open access article distributed under the terms and conditions of the Creative Commons Attribution (CC BY) license (https://creativecommons.org/licenses/by/4.0/).

implementing practical stream summaries without the need to access the slower external DRAM. Third, the extensive Intellectual Property (IP) support and flexible high-bandwidth Input/Output (I/O) in FPGAs allow for the easy integration of FPGAs in edge and cloud applications that deploy data stream analytics.

With the advances in High-Level Synthesis (HLS) tools, functional hardware implementations of software algorithms can be easily and quickly realized. However, design optimizations are often necessary to achieve the optimal operation of hardware accelerators. The main goal in HLS design is to be able to implement a stall-free accelerator or, in other words, an accelerator with an Initiation Interval (II) of 1. An II of 1 means that the accelerator pipeline can process a new input item every clock cycle without stalling. An optimal stall-free accelerator is not always possible, especially if the accelerator requires complex memory accesses. For example, a hardware implementation of a hash table will not maintain an II of 1 due to the pipeline stalls needed to resolve collisions when inserting items in the table.

There are several complexities associated with designing an optimal accelerator to compute the heavy hitters in data streams, depending on the base algorithm used and the overall system requirements. For example, several heavy hitter algorithms utilize hash table data structures for counting item occurrences in a data stream, preventing a stall-free operation due to hash collisions and memory dependency. Additionally, if the accelerator is required to support the top-k item query, this adds further complexity to the design. A suitable data structure, such as a priority queue, needs to be implemented to maintain a list of the top k items. In addition, a suitable interface is needed for the host to traverse the top-k item list. Several hardware architectures have been proposed in the literature to accelerate the heavy hitter problem in data streams (See Section 2). However, there is no single elegant solution to handle the aforementioned complexities without scaling down the design. This paper presents a novel hardware adaptation of the approximate *Probabilistic* sampling algorithm in [8], which is used for the top-k item query in data streams. The presented hardware architecture aims to address design complexities in existing solutions by introducing hardware-specific optimizations at the algorithmic level. These modifications are based on intuition and favor simplicity and design scalability to facilitate the strict mapping of the algorithm into hardware. When implemented as an HLS kernel using Intel HLS compiler and targeting an Intel Stratix 10 FPGA, the proposed architecture scaled very well and achieved higher throughputs compared to all relevant existing FPGA accelerators. Furthermore, test results on synthetic and real datasets showed near-perfect accuracy—exceeding 95% in all test runs. This is a significant improvement over previously proposed scaled down accelerators that would strictly map the *Probabilistic* sampling algorithm into hardware. In short, the main contributions in this paper can be summarized as follows:

(1) A novel hardware-optimized algorithm for computing the top-k query in data streams. The algorithm is the first to deploy techniques such as fingerprinting, optimistic counting, re-hashing, and timestamping to resolve hardware-specific complexities usually associated with relevant FPGA accelerators.
(2) An HLS kernel design for the proposed optimized algorithm that can be easily reproduced using HLS tools from both major FPGA vendors (AMD/Xilinx and Intel). The HLS kernel also deploys novel optimizations at the hardware level to resolve common implementation issues such as memory dependency and data hazards.
(3) The fastest FPGA implementation compared to existing accelerators, achieving high throughputs even when the implementation has a high chip utilization.
(4) Addressing important practicality issues in kernel design such as larger key sizes (up to 128-bit), result mergeability, and parallelism.

The remainder of this paper is organized as follows: Section 2 briefly introduces the top-k query problem in data streams and discusses some of the most relevant previously published work. Section 3 briefly discusses the *Probabilistic* sampling algorithm, which is used as the basis for the top-k item query computation. Section 4 presents the proposed optimizations for efficient hardware implementation. Sections 5 and 6 details the HLS accelerator kernel design and FPGA implementation. Section 7 presents the functional verification and evaluation of the proposed accelerator. Finally, conclusions and future work plans are summarized in Section 8.

2. Background and Related Work

2.1. The Top-k Item Query in Data Streams

Assume we have a stream *S* of size *N*, and we want to find the *k* most frequent items in *S* (known as top-k items or top-k heavy hitters). The size of the stream *N* is not necessarily known beforehand. A naïve exact solution for finding the top-k items can be realized using a lookup table data structure with key-count pairs in its entries. For every item hit, if the item key is in the table, its count value is updated. Otherwise, a new entry is created in the table. An additional priority queue data structure can be used to maintain a record of the current top-k items. Alternatively, the entries in the lookup table can be sorted according to their count values to extract the top-*k* items when the stream is exhausted or when results are required. For a general input distribution containing a large number of distinct items, finding an exact solution is impractical or even impossible due to the time and space complexity.

In most practical applications, an exact result is not required. Approximate results can be obtained using approximate algorithms that do not count every distinct item in the stream. In general, approximate algorithms only maintain a summary of the stream and a list of heavy hitter candidates that most likely contain most of the actual top-k items in the stream. There are two categories of such approximate algorithms in the literature: counter-based and sketch-based algorithms.

2.2. FPGA Implementations of Counter-Based Algorithms

In general, counter-based algorithms allocate counters in memory, only enough for counting a small subset of the overall distinct items in the stream. Each counter is a key-value tuple, where the key is an identifier for a heavy hitter candidate and the value is the estimated count for this candidate. When the stream is processed, the counters should contain all or most of the heavy hitters in the stream. Counter-based algorithms do not require an additional priority queue for computing the top-k items, as this can be performed by using a simple sort operation.

Several architectures have been proposed to accelerate item counting using FPGAs. Early works on the related field of itemset mining acceleration with FPGAs proposed implementing fine-grained systolic arrays with serially connected Processing Elements (PEs). Each PE contains a small independent memory allocated for updating a single or small number of key-value tuples [9]. A limited number of such systolic array accelerators have been specifically designed to compute the heavy hitters in data streams [10–13]. Most of these accelerators implement the popular *Space-Saving* algorithm [14]. *Space-Saving* uses *m* counters to monitor the first *m* distinct items that appear in the stream. A new incoming item, not in the any of the *m* counters, replaces the item with the minimum count. By doing so, frequent items with large counts should remain in some of the counters when the stream is exhausted. Although implementing *Space-Saving* is relatively simple in software, mapping it to a systolic array architecture can be tricky with complexities that limit the overall number of monitored items. Current *Space-Saving* FPGA implementations can be used to monitor hundreds to a few thousands of items using mid-capacity and large-capacity FPGA chips [11–13]. The work in [10] showed that the *Probabilistic* sampling algorithm proposed in [8] maps better to a systolic array architecture, resulting in some notable improvements compared to *Space-Saving*.

All the aforementioned systolic array accelerators rely solely on the FPGA logic resources for implementing small distributed memories to store the key-value tuples. Although they are stall-free and relatively fast, the maximum number of items that can be monitored is limited, especially with larger key integer sizes. Several works have demonstrated that it is possible to expand the total number of monitored items using key-value store approaches based on hash tables that utilize the abundant embedded memory resources on an FPGA chip [15,16]. However, the performance of such approaches significantly suffers because of hash collisions and the pipeline stalls needed for updating the monitored items.

2.3. FPGA Implementations of Sketch-Based Algorithms

Sketch-based algorithms aim to summarize the frequency distribution of the data stream using a unique data structure referred to as a "sketch". The frequency estimation sketch can be queried to output the estimated count of a particular item key. A very popular frequency estimation sketch is the *count–min* sketch [17]. The *count–min* sketch consists of several hash tables with different hash functions. The hash tables only contain count values in their entries. When the sketch is updated with a new item hit from the stream, the item key is hashed into all of the tables, and the relevant table entries are incremented. Since the hash functions are different, item collisions will differ in all of the tables. For any queried item, the table entry with the minimum count (minimum number of collisions) represents the best count estimate for the item.

The count–min update process does not require hash collision resolution. Several FPFA accelerators implement *count–min* on the on-chip embedded memory to realize constant update time and, in some cases, stall-free operation [18–23]. As the *count–min* does not store item keys in its table entries, it cannot be directly used to solve the top-k item query problem. An additional hardware data structure is required to maintain a record of the top key-value tuples [24–26]. Only a few FPGA implementations extend the basic functionality of *count–min* to support the top-k query. The sketch accelerator in [27] uses a simple priority queue architecture. Because the queue update process is sequential, the throughput can be drastically reduced, even for small values of k. A more sophisticated and improved priority queue was later presented in the accelerator proposed in [28,29]. This accelerator uses a large portion of the available embedded memory resources for implementing the queue rather than for the sketch. The queue is implemented as a pipelined hash table with several independent buckets to allow for consecutive updates (1 update per clock cycle in most cases). This allows one to monitor a larger number of top items, but with reduced count accuracy due to the smaller sketch. In fact, the main objective of this accelerator was not to output the top-k items accurately but to estimate the entropy of the input stream using the top-k item list as a sample of the stream.

A related work deploys a hybrid approach, combining a sketch implemented on embedded memory and a systolic array implemented using the logic resources of the FPGA. The sketch is only used as a filter that passes item hits for items with counts larger than a specific threshold to the systolic array that monitors the heavy hitters [10]. While this architecture achieved good performance, it only allowed one to monitor a relatively small number of items. Another related work implements an accelerator based on an alternative sketch algorithm that supports the top-k query without the need for a priority queue [30]. The implemented *Heavy-Keeper* sketch uses hash tables similar to *count–min*; however, item keys are also stored in the table entries, allowing for the sketch to be traversed to extract the top-k items [31]. Due to the complexity in updating the *Heavy-Keeper* sketch, the item update process required several clock cycles (II larger than 1).

2.4. Summary of Existing FPGA Implementations

Table 1 summarizes the most relevant existing FPGA implementations of data stream heavy-hitter detection algorithms. Implementations are classified into three categories: systolic array, hash table, and sketch implementations. These categories are compared according to the following attributes:

Design: A stalling design means that the accelerator cannot guarantee that an item is processed in a single clock cycle, resulting in a slow FPGA implementation.
Key Size: Smaller key sizes limit the possible applications.

Top-k Query Support: Several implementations do not directly support the top-k query as an additional priority queue is needed.
Result Readout Style: Using a buffer implemented as a memory block allows the results to be easily copied and processed by a host. However, a large buffer increases latency and host processing time. A First-In First-Out (FIFO) interface is present in accelerators that are only capable of outputting results for individual item count queries.

Table 1. Summary of FPGA implementations of data stream heavy-hitter detection algorithms.

Ref.	Type	Algorithm	Stalling Design?	Key Size (Bits)	Top-k Query?	Result Readout
[10]	Systolic Array	Probabilistic [8]	No	32	Yes	Small Buffer
[11–13]	Systolic Array	Space-Saving [14]	Yes	32	No	FIFO
[15,16]	Hash Table	Cukoo Hash [32]	Yes	104	Yes	Large Buffer
[18–23]	Sketch	Count–Min [17]	No	32-128	No	FIFO
[30]	Sketch	Heavy-Keeper [31]	Yes	32	Yes	Large Buffer

From Table 1 we can see that there is no single FPGA implementation that excels in all attributes. Therefore, the presented work focuses on filling this gap by addressing all the attributes needed to realize a fast and practical FPGA accelerator specifically tailored for the top-k query in data streams.

3. Base Algorithm: Probabilistic Sampling

Our proposed approach for finding the top-k items borrows several ideas from the *Probabilistic* sampling algorithm in [8]. We first briefly introduce *Probabilistic* sampling in this section and, later, we detail the proposed hardware-specific optimizations in Section 4.

Probabilistic sampling is generally considered fast and efficient, and can approximate a list of the top-k items in data streams. The idea behind *Probabilistic* sampling is very simple and as follows: the data stream is divided into rounds of size r that are processed separately. The algorithm uses m counters to count the first m distinct items that appear in the round (see Figure 1). Hits from items not registered in the m counters are discarded. A hash table with key-value entries is a fast and simple method for counting the sampled items. At the end of a round, the k items with the largest round counts are extracted and stored in a list. The process is then repeated in later rounds of the stream. The final list of approximate top-k items is obtained by merging the top-k lists at the end of each round. Only the items with the highest round counts make it to the final list. For duplicate item records, only one entry with the highest round count is stored in the merged list.

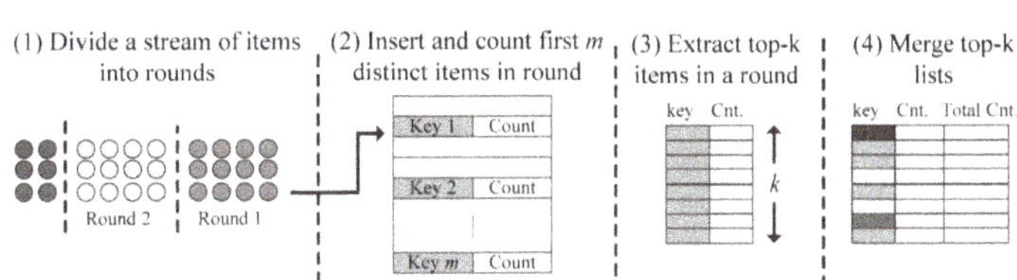

Figure 1. Probabilistic sampling algorithm.

In addition to the highest round count for an item in the top-k list, an extra total accumulate count field can also be stored for each item. This field represents an underestimated value for the item count. The value in this field is obtained by accumulating the round counts of items appearing in the top-k list in consecutive rounds. If an item is recorded as a top item in all rounds, then the total count estimate is the actual count of this item.

4. Proposed Approach: Optimizations for Efficient Hardware Implementation

4.1. Updating the Round Table

To implement the *Probabilistic* algorithm on an FPGA, a hash table is needed for counting items in a round. The hash table can be implemented as a RAM block. The aim is to store as many key-value pairs as possible using the available on-chip memory. In addition, a stall-free operation is necessary to achieve high throughput. The main obstacle in achieving a stall-free operation is the item collision issue. There are some trivial methods to minimize hash collisions. For example, we can increase the load factor of the hash table by using a RAM block with M buckets to count m distinct items ($M > m$). Additionally, it is possible to use an additional bloom filter data structure to decrease the number of table updates [33]. However, using these methods will increase the memory usage and limit the number of items that can be monitored in a round.

Optimizing memory accesses and the RAM block geometry can also enhance the performance of a hash table. For example, breaking the RAM block into several pipelined banks allows for several concurrent memory accesses [16,33]. However, real data streams are typically skewed, causing contention on some of the memory banks and, as a result, preventing meaningful performance gains. While there are some methods that have been proposed to reduce contention on the memory banks when processing has skewed streams [34,35], there is no efficient solution that guarantees a stall-free operation.

Instead of strictly mapping the round table update process in *Probabilistic* models into hardware, we modify this process to address the aforementioned shortcomings associated with hardware hash tables. The remainder of this section details the proposed optimizations for the round table update process. Figure 2 shows the pseudo code for the optimized *update_table()* function, which returns the round count for a particular item hit in a round. The function has four arguments:

(1) FP: item fingerprint.
(2) TS: timestamp.
(3) IDX: table index generated by a hash function.
(4) w: weight of an item ($w = 1$).

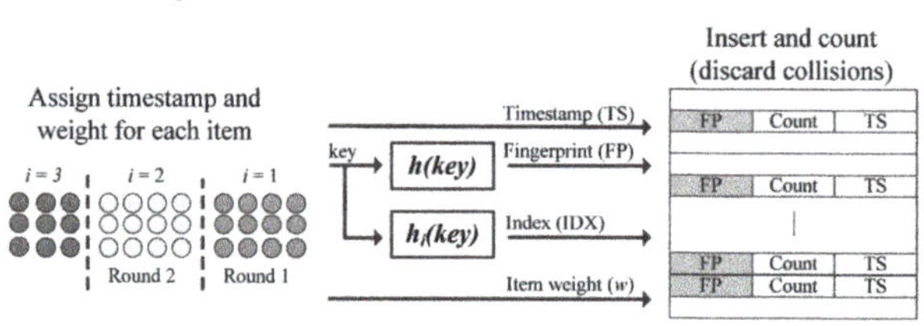


```
Function update_table(FP, TS, IDX, w):
    c ← 0
    if table[IDX].TS ≠ TS then
        table[IDX].FP ← FP
        table[IDX].TS ← TS
        table[IDX].count ← w
        c ← w
    else if table[IDX].FP = FP then
        table[IDX].count ← table[IDX].count + w
        c ← table[IDX].count
    return c
```

Figure 2. The proposed optimized function for updating the round table.

4.1.1. Optimistic Counting

As mentioned earlier, the main obstacle in achieving a stall-free operation is item collisions when counting the first m distinct items in a round. Rather than increasing the hash table load factor to decrease item collisions, we opt for an "optimistic counting" approach that ignores item collisions. The idea behind optimistic counting is simple and can be described in four steps as follows:

(1) An item hit is hashed to generate an index (IDX) to a bucket in the table.
(2) If the indexed bucket is empty, create an entry for the item in the bucket.
(3) If the indexed bucket already contains an entry for the item, increment the round count of the item.
(4) If the indexed bucket contains an entry for another item, discard the item hit.

We can see that the optimistic counting approach utilizes all M buckets for item counting opposite to the conventional hash table approach, which only inserts m items in the M buckets ($M > m$). The first item to be hashed into an empty bucket will stick in the bucket for the remainder of the round. Item collisions are totally ignored in favor of sampling more items and achieving a stall-free operation in hardware. We refer to the optimized function as "optimistic" because it assumes, or in other words, hopes that there will be no item collisions before, at least, the m items are inserted into the table. Due to the simplicity of this approach, we can count a significantly larger number of items using the same amount of FPGA embedded memory compared to a conventional hash table with a large load factor. In addition, to allow for a stall-free operation, we can also argue that the proposed simplified item counting technique can result in better accuracy compared to a hash table with a large load factor when the amount of embedded memory is limited. This is mainly attributed to the larger number of sampled items using the same amount of memory.

4.1.2. Fingerprinting

As the on-chip memory in FPGAs is generally limited, it is very important to optimize memory usage to be able to sample as many items as possible. Most of the available accelerators discussed in Section 3 limit the key size of an item to 4 bytes. While this is sufficient for some applications (example: IP address in IPv4), there are other applications that require larger key sizes (example: 128-bit IP address in IPv6). Storing the full item keys in the round table buckets will limit the total number of buckets possible with the available memory, especially for larger key sizes. As we are only interested in maintaining a record of the top-k items, there is no need to store the keys for all the sampled items. Alternatively, we can store a unique fingerprint (FP) of the item in the round table [31]. This fingerprint is generated by a hash function and is much smaller in size compared to the actual key (see Figure 2). Simply, when calling the optimized *table_update()* function, the function matches the generated item fingerprint to the fingerprint stored in the indexed table bucket to decide if the round count should be incremented.

4.1.3. Round Re-Hashing

As our simplified optimistic counting technique ignores item collisions, we need to consider the case when two or more heavy hitter items generate the same index early in the round. Only one of these heavy hitters will be registered in a round and considered as a top-k candidate. In addition, there is a possibility of different items generating the same table index as well as the same fingerprint.

To rectify the issue of colliding heavy hitters, we propose round re-hashing. Basically, the seed for the hash function used to generate the item index is randomly generated for each round (labelled $h_i(key)$ in Figure 2). The idea behind round re-hashing is simple and as follows: if two heavy hitter items collide in a round, it is highly unlikely that the same items will collide with each other again in another round, as they will likely generate different hashes.

4.1.4. Using a Timestamp for Reduced Latency

Another issue that needs to be addressed when updating the round table is the RAM block reset needed in between rounds. As we are going to use the same RAM block for counting items in different rounds, all buckets must be reset before starting a new round. This is time-consuming, as the memory locations in the RAM block need to be reset sequentially. The latency of the accelerator will significantly increase, especially if the RAM block is large and the chosen round size is not sufficiently large relative to the RAM block size.

To address this issue, we propose using a unique timestamp for every round. Items in the same round will be assigned the same timestamp. When an item is first registered in a table bucket, the current timestamp is also stored in the bucket (see Figure 2). When updating the table, if an indexed bucket has a timestamp different to the current timestamp, the bucket is considered empty and can be used to register a new item. This way, there is no need to reset the RAM block after each round. To avoid significantly increasing the memory usage of the round table, only small numbers should be used for the timestamp. For example, if 1 byte is allocated for the timestamp, this allows one to run 255 rounds before a reset of the RAM block is required.

4.2. Updating the Heavy Hitter Summary

After counting items in a round, the top-k frequent items in the round should be extracted and stored in a list. In hardware, the process of updating the top-k item list should run concurrently with the table update process to achieve a stall-free operation. Typically, a priority queue data structure is used to maintain a record of the top-k items; however, FPGA implementations of such data structures can be complex and inefficient, especially if a stall-free operation is required (see Section 2).

We propose a data structure that is entirely different to a priority queue. We call this data structure the "heavy hitter summary". The summary is a hash table with K buckets, where K is much larger than k but still much smaller than M. Figure 3 shows the pseudo code for the *update_summary()* function, which is called after the *update_table()* function in Figure 2.

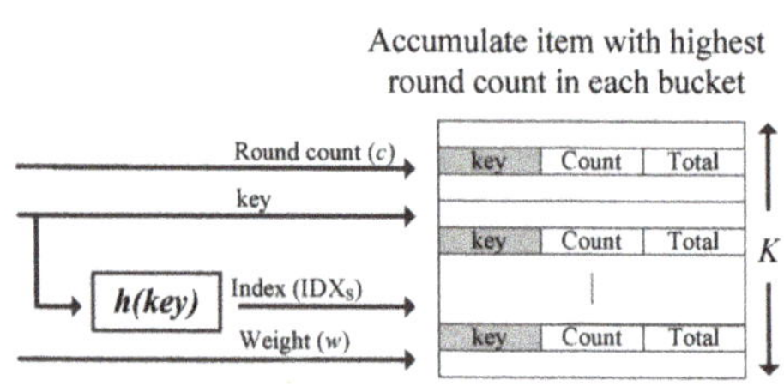

Function update_summary(key, IDX_S, c, w):
  if $summary[IDX_S].key = key$ **then**
    $summary[IDX_S].total \leftarrow summary[IDX_S].total + w$
  if $summary[IDX_S].count < c$ **then**
    $summary[IDX_S].count \leftarrow c$
    if $summary[IDX_S].key \neq key$ **then**
      $summary[IDX_S].key \leftarrow key$
      $summary[IDX_S].total \leftarrow w$

Figure 3. Updating the heavy hitter summary.

Each bucket in the summary contains three fields: an item key, round count, and total accumulate count for the item. The *update_summary()* function arguments are the key for the current item hit, the bucket index (IDX_S) generated by the hash function, the item's round count returned by the *update_table()* function, and the weight of the item. The procedure for updating the summary can be summarized as follows:

(1) An item hit is hashed to generate an index (IDX_S) to a bucket in the summary.
(2) If the key matches the key stored in the indexed bucket, the accumulate count is incremented. The stored round count is also updated in case the current round count is larger than the stored count.
(3) If the key is different than the stored key in the indexed bucket, the bucket is updated with the new item only if the new item round count is larger than the stored round count.

The *update_summary()* function is called for every item hit in the stream. The summary is only reset when the stream is exhausted. The function basically splits the stream into K sub-streams using the hash function. Only the heaviest item from each sub-stream is maintained in the relevant bucket (the item with the largest round count). When K is sufficiently large relative to k, most of the actual top-k item should stick in some of the summary's buckets. Querying the top-k items from the summary is simple and involves the following: first the items in the summary are sorted according to their accumulate count. Then, the top-k items are extracted. The latency incurred from the sorting operation should not affect the performance when implementing the proposed summary in hardware. This is because, in practical applications, a top-k query is only issued intermittently after very large intervals of streaming activity. Additionally, K is generally small, and the sorting operation can be efficiently completed in software by a host.

5. HLS Kernel Architecture

This section details the design of an FPGA accelerator implementing the proposed algorithm in Section 4. We draw the readers' attention to the Intel HLS documentation [36], as the remainder of this section uses technical terminology that may be specific to Intel HLS design flow. As the presented design only deploys standard pragmas, the design can be easily migrated to other HLS tools (for example, AMD/Xilinx Vitis HLS). In addition, with minor modifications, the accelerator can be deployed as a kernel in Intel heterogeneous computing tools such as the following: Intel FPGA SDK for OpenCL and Intel OneAPI toolkit. Since these tools use the same core compiler technology as Intel HLS, results should be the same regardless of which Intel design tool is used.

5.1. Architecture Overview

The accelerator is designed as a kernel that operates alongside a host CPU, which is the typical hardware setup in streaming applications. The kernel is implemented as a slave component that is controlled by a host through a memory mapped slave interface (see Figure 4). In Intel HLS compiler, the "hls_avalon_slave_component" attribute can be used to infer a slave interface compatible with the Avalon bus specification. The host launches the kernel to process a single round of the stream. There are some memory mapped registers that need to be setup by the host before launching the kernel, including the following: the round size (r), the timestamp of the round (TS), and the hash function seed needed for generating the round table index (IDX), as explained in Section 4. The kernel contains two memory blocks, one represents the round table with M memory locations and the other represents the heavy hitter summary with K memory locations. Both memory blocks are implemented as simple dual-port RAM using the M20K embedded memory resources in Intel FPGAs. A simple dual-port memory has a read port dedicated to reading (load operations) and a write port dedicated to writing (store operations). Using the relevant component macros in Intel HLS compiler, the heavy hitter summary is specified as a slave memory with read access granted to the host. By doing so, the compiler inserts arbitration logic at the read port of the heavy hitter summary to allow the host to read the summary when results are required. For a simpler arbitration logic, we prevent host access during a round when the kernel is active.

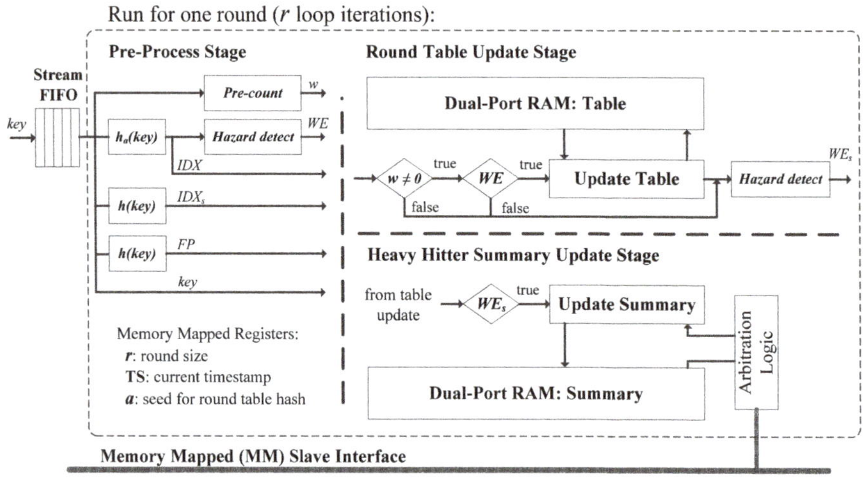

Figure 4. Architecture overview of the HLS kernel.

After launching the kernel, the kernel runs for r iterations. In every iteration a blocking read operation from an input stream FIFO is performed to read the item keys from the data stream. The kernel pipeline can be divided into three main stages: the pre-process stage, the round table update stage, and the heavy hitter summary update stage. In the pre-process stage, the input keys are processed to generate the hashes required in the proposed algorithm. In addition, some control signals are generated to efficiently handle data hazards in later stages in the pipeline. The round table update stage basically executes the function in Figure 2, while the heavy hitter summary update stage executes the function in Figure 3. While a functional hardware implementation of the proposed algorithm in Section 4 is very straightforward using HLS, achieving an II of 1 requires some design optimizations, which will be detailed in the remainder of this section.

5.2. Round Table Load-Store Logic

A naïve HLS code for the *update_table()* function in Figure 2 will certainly result in a pipeline with an II larger than 1 when compiled. The pipeline will not be able to process an input item every clock cycle mainly due to memory dependency and the read-modify-write operation performed when updating a bucket in the round table. With dual-port RAM, it is possible to perform load and store operations at the same time; however, the minimum latency of a RAM block is 1 clock cycle. This means that consecutive updates to the same bucket will create a data hazard issue. To prevent functional failure due to memory dependency, the HLS compiler inserts stalling logic in the pipeline and increases II to a number larger than the RAM block latency. Therefore, the best possible II for a naïve HLS implementation is 2. In addition, the performance will further suffer if the round table is large—consisting of many FPGA RAM primitives that are physically distanced apart on the chip. The compiler may decide to further increase II or reduce the maximum operating frequency (fmax) to meet timing requirements.

HLS tools usually support special pragmas to relax memory dependency. For example, the "ivdep safelen (m)" pragma can be used to tell the compiler that there will be no memory dependency for at least m loop iterations. We only need $m = 2$ to achieve an II of 1. However, when m is sufficiently larger than the RAM block latency, the compiler will be able to schedule the memory load and store operations further apart in the pipeline to achieve higher fmax, and this is particularly useful when the RAM block is large [37].

Relaxing memory dependencies and specifying a safe dependence distance m for the compiler does not guarantee functional correctness. The programmer needs a mechanism to guarantee that no such dependencies will occur in the first place to prevent functional failure. There are several solutions available in the literature for handling dependencies at run-time when updating frequency estimation sketches (Examples: [19,21,37,38]). None of the available solutions are directly applicable to our proposed algorithm. Therefore, we propose a custom solution for handling memory dependencies when updating the round table according to the *update_table()* function in Figure 2. The solution is based on a Load-Store Queue (LSQ) that precedes the round table update stage in Figure 4. The LSQ mainly performs two tasks—labelled "pre-count" and "hazard detect" in Figure 4.

5.2.1. Pre-Count: Forward Weight Accumulation

Memories constructed using the logic blocks of the FPGA do not have read and write latencies as in memories constructed with the embedded RAM resources of the FPGA. The first step in our proposed solution for handling memory dependencies is to pre-count the occurrences of item keys in a small buffer constructed using the logic blocks. We refer to this technique as "forward weight accumulation". The circuit used for forward weight accumulation is shown in Figure 5. The circuit consists of a shift register of size m. All registers have parallel connections to the weight accumulation logic.

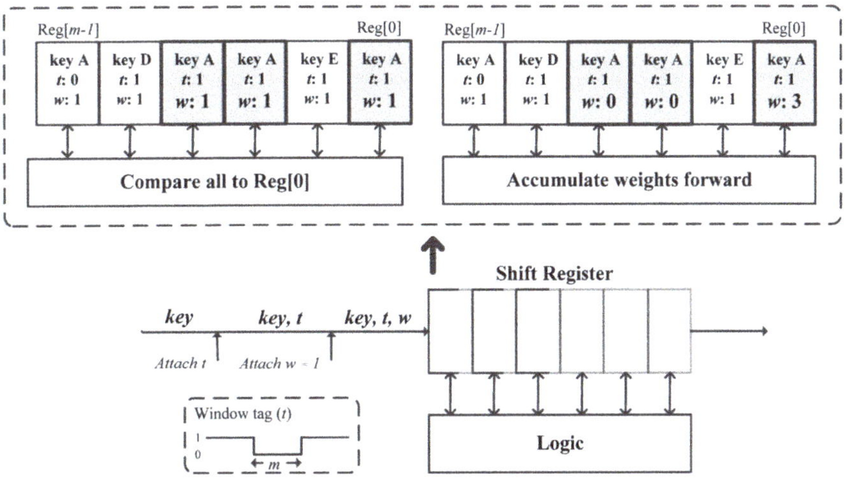

HLS Code Snippet

```
#pragma unroll
for (int k = (m - 1); k > 0; --k) {
    if ((shift_reg[0].key == shift_reg[k].key) && (shift_reg[0].t == shift_reg[k].t)) {
        shift_reg[0].w += shift_reg[k].w;
        shift_reg[k].w = 0;
    }
    }
```

Figure 5. Pre-count: breaking memory dependency for m loop iterations using forward weight accumulation.

When the kernel is launched, the item keys are read from the input FIFO. The kernel will first wrap each item key in a data structure containing a wight variable (w), which is initialized to 1 (see Figure 5). In addition to the key and initial weight of 1, the kernel will wrap a variable t that is assigned to either 0 or 1 depending on the order of the item in the stream. The variable t is a tag used for dividing consecutive keys into groups or windows of size m; each key in the same window is assigned the same tag. Basically, the value of t is inverted in every m loop iteration. When passing items through the shift register, the weight accumulation logic will compare the key in Reg[0] to the keys in all other registers. The weight of any identical key belonging to the same window will be accumulated forward in Reg[0] before it is reset to zero. Any key with a weight of zero is regarded as a "bubble" and can be discarded in later stages in the pipeline when it exits the shift register. By inserting bubbles in the pipeline, memory dependencies are eliminated for at least m loop iterations.

Implementing forward weight accumulation is straightforward with HLS as it only requires a simple loop unroll pragma (see the HLS code snippet in Figure 5). It should be noted that a previously proposed solution deployed a similar backward weight accumulation technique that does not require the stream to be divided into windows with alternating tags [37]. The idea was to keep accumulating wights backward in Reg[$m-1$] to insert as many bubbles as possible into the pipeline. While the solution is perfect for simple frequency estimation sketches, it is not suitable for the algorithm presented in this paper. This is mainly because, in skewed data streams, the weights of frequent items will likely keep accumulating in the shift register for long intervals in the case of using a backward accumulation technique. This is highly undesirable for our sampling algorithm because updates of frequent items are delayed in the shift register and may fail to stick in any of the round table buckets when they eventually exit the shift register.

5.2.2. Data Hazard Detection

With forward weight accumulation, we resolve the memory dependency issue associated with consecutive items with identical keys in the stream. There is still one minor issue that needs to be resolved before safely using the "ivdep safelen (m)" pragma in the kernel's HLS code. The issue arises from the fact that different keys may generate the same round table index (IDX). While this is not a problem when the keys are distanced apart in the stream or when a key is already registered in the indexed bucket, a data hazard occurs when two keys that are less than m cycles apart generate the same index to an empty bucket. When the empty bucket is evaluated for the first key update, a memory store operation is initiated to register the key in the bucket. Due to memory latency, the bucket may still be interpreted as empty when evaluated for the second key update—initiating an incorrect memory store operation.

To resolve this data hazard, the table indexes are first pre-processed by an LSQ similar to the one used for forward weight accumulation. The LSQ consists of a shift register of size m and some data hazard detection logic (see Figure 6). When passing the indexes to the shift register, the kernel wraps a Write Enable (WE) variable as well as the same window tag t used in forward weight accumulation. The data hazard detection logic compares the index in Reg[$m-1$] to all the indexes in the other registers. If any identical index that belongs to the same window is detected, the WE variable in Reg[$m-1$] is reset. In later stages in the pipeline, any key update with WE = 0 is discarded.

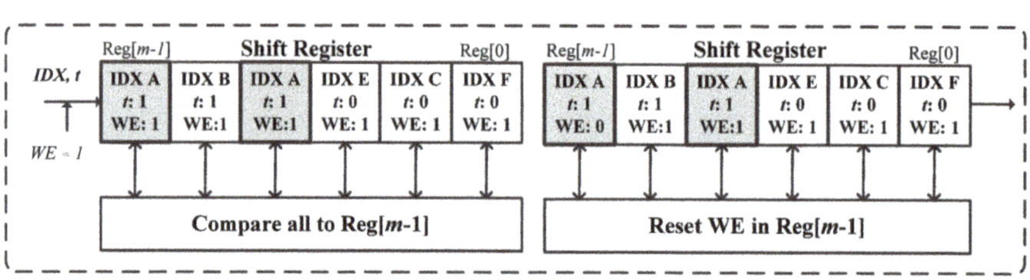

HLS Code Snippet
```
#pragma unroll
    for (int k = 0; k < (m - 1); ++k) {
        if ((shift_reg[m - 1].IDX == shift_reg[k].IDX) && (shift_reg[m-1].t == shift_reg[k].t)) {
            shift_reg[m - 1].WE = 0;
        }
    }
```

Figure 6. Resolving data hazards before the round table update stage.

5.3. Heavy Hitter Summary Load-Store Logic

The next stage in the pipeline is the heavy hitter summary update stage, which will execute the *update_summary()* function in Figure 3. This function takes the item key, summary index (IDX_S), the item weight w, and the round count c.

As memory dependencies occur due to identical keys in the stream that are already resolved by weight accumulation, there is only one minor data hazard that needs to be resolved before executing the *update_summary()* function. The data hazard occurs when two item hits with different keys are less than m cycles apart and generate the same summary index. In particular, if the indexed bucket stores a round count smaller than the round count of both items and the round count of the second key is larger than the round count of the first key. When the bucket is evaluated for the first item, a store operation will be initiated to replace the stored item with the smaller round count. Due to memory latency, an incorrect memory store operation may be also initiated for the second item. To resolve this data hazard, another LSQ is used after the round table update stage. The LSQ in Figure 7 consists of a shift register of size m and some data hazard detection logic. The summary indexes for the items are wrapped with the associated round counts as well as a Write Enable (WE_S) variable before being passed to the shift register of the LSQ. The data hazard detection logic compares the index in Reg[$m-1$] to all the indexes in the other registers. If any identical index with a larger round count is present, the WE_S signal in Reg[$m-1$] is reset. Any item with $WE_S = 0$ is discarded before updating the heavy hitter summary.

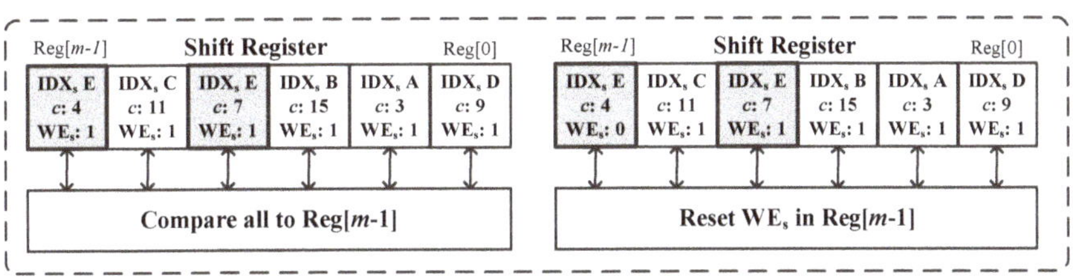

HLS Code Snippet

```
#pragma unroll
for (int k = 0; k < (m-1); ++k) {
    if ((shift_reg[m-1].IDXs == shift_reg[k].IDXs)&&(shift_reg[m-1].c < shift_reg[k].c))
    {
        shift_reg[m-1].WEs = 0;
    }
}
```

Figure 7. Resolving data hazards before the heavy hitter summary update stage.

By applying the aforementioned LSQs, a stall-free HLS kernel for the proposed algorithm can be easily implemented. The kernel can be invoked by a host to process as many rounds as needed. However, an additional minor tweak is needed to safely process multiple rounds. The shift registers of the LSQs need to be flushed at the end of a round to ensure that all item hits are processed. To accomplish this, the kernel runs for a number of extra iterations at the end of the round. During these iterations, the kernel passes dummy item keys with zero weights to flush the shift registers before the next round.

5.4. Support for Parallel Sub-Streams

Achieving high performance when mapping algorithms into hardware can be performed via deep pipelining and task parallelism. Task parallelism involves vectorizing the input into distinct sets that are processed using separate kernels. Results from these separate kernels can then be merged. By applying task parallelism, the throughput can be significantly increased without the need to operate a kernel hardware at its fmax. Task parallelism is particularly useful to exploit the high throughputs supported by the on-chip

High Bandwidth Memory (HBM) available in some of the trending high-end FPGAs [39]. In addition, FPGAs with high-speed transceivers may require several replicas of a kernel to be able to saturate the available bandwidth.

In order to apply task parallelism, the algorithm should support vectorization. In general, data stream frequency estimation sketches, such as the *count–min* sketch, can be vectorized by splitting the stream into several sub-streams that update separate sketches with identical geometry. Merging results from distributed *count–min* sketches with identical geometry is simple as it only involves the entry-wise summarization of the sketch tables. In hardware, this can be difficult, as the host needs to access every table entry in the sketch, which can span most of the embedded RAM on the FPGA chip. Based on the lack of attempts reported in the literature, there have been few attempts to implement parallel frequency estimation sketches in a single FPGA chip [19,21]. These parallel systems only support a general update–query model, where the frequency of individual items is sequentially acquired by the host. This simple model does not support the top-k item query as it does not allow the host to access the sketch memory or any priority queue paired with the sketch.

One of the important advantages of the algorithm proposed in this paper is the ease of vectorizing the algorithm in hardware. First, the data stream is naturally divided into rounds that can be processed independently by several kernel replicas (see Figure 8). Second, the host only needs to access the smaller heavy hitter summary data structure in each kernel, which is implemented as a slave memory. Third, the merging of heavy hitter summaries is very simple, especially if the same hash function is used for all the summaries. The pseudo code in Figure 8 shows how two summaries from two kernel replicas can be merged when the same hash function and the same round sizes are used in the different kernel replicas. The merge process can be performed efficiently by the host as it only has a time complexity of O(n), where ($n = K$). In addition, if the heavy hitter summary configuration is changed in the HLS code to grant both read and write access to the host, the merge process can be performed in-place without the need to copy the summaries to the host's memory.

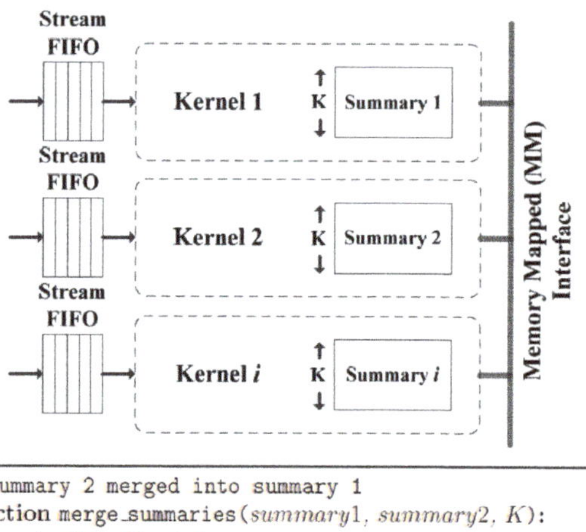


```
// summary 2 merged into summary 1
Function merge_summaries(summary1, summary2, K):
    for i = 0 to K − 1 do
        if summary1[i].key = summary2[i].key then
            summary1[i].total ← summary1[i].total + summary2[i].total
            if summary1[i].count < summary2[i].count then
                summary1[i].count ← summary2[i].count
        else if summary1[i].count < summary2[i].count then
            summary1[i] ← summary2[i]
```

Figure 8. Scaling performance with parallelism.

6. FPGA Implementation

This section reports the FPGA implementation results when compiling the proposed kernel using Intel HLS compiler and Intel Quartus Prime synthesis tools. The midrange Arria 10 GX 1150 FPGA was selected as a target device. This is the largest chip from the Arria 10 family, with 427,200 Adaptive Logic Modules (ALMs), 1518 Digital Signal Processing (DSP) blocks, and 2713 M20K embedded RAM blocks. As the kernel mainly consumes embedded RAM for implementing the round table and the heavy hitter summary, we focus on validating the notion that the throughput of the kernel is sustained even when scaling the size of the RAM blocks to large portions of the available on-chip memory.

As there are many parameters that can be varied in the kernel's configuration, we fix some of these parameters to practical values for simpler analysis. First, the size of the heavy hitter summary K is fixed to 2^{15} in all configurations. This size should be selected according to the number of top-k items that needs to be monitored. Making K unnecessarily large may increase the time for result readout without gaining meaningful improvements in accuracy. So, if we are aiming to report somewhere near the top-1000 items, a fixed size of K = 2^{15} is a reasonable choice that should provide a good balance between accuracy and result readout time. The size of the round count variable c is fixed to 16-bit, the total accumulate count filed in the heavy hitter summary's buckets to 32-bit, the item fingerprint FP to 16-bit, and finally, the round timestamp TS to 8-bit.

As can be seen from Figure 4, there are three hash circuits that need to be implemented in the kernel. The quality of the hash function used will affect the accuracy of the system. We opt for using a simple and efficient hash function that can be implemented with the least amount of FPGA resources. We use the "binary multiplicative" hash function described and analyzed in [40] for the three hash circuits in the kernel. This hash function only requires a single multiplier when implemented in hardware. The operation of the hash function is described as follows: assume we have L-bit integers that need to be mapped to l-bit integers. Using an odd integer seed a, the hash function is defined as:

$$h_a(x) = (a.x) mod\ 2^L / 2^{L-l}$$

Table 2 reports the FPGA post place-and-route implementation results for nine different configurations of the kernel. In these configurations, the size of the round table M and the size of the item key are varied. Three sizes of M were considered: 2^{17}, 2^{18}, and 2^{19}, and three key sizes were considered: 32-bit, 64-bit, and 128-bit. In all configurations, the safe dependence distance m was fixed to 8. Additionally, a target fmax of 400 MHz and a target II of 1 were specified to the compiler when compiling all configurations.

Table 2. Implementation results on Intel Arria 10 GX 1150 FPGA.

M	Key Size (Bits)	Fmax (MHz)	Resource Utilization		
			ALM	DSP	M20K
2^{17}	32	417	1577	6	480 (18%)
	64	399	2294	18	544 (20%)
	128	379	3493	45	672 (25%)
2^{18}	32	351	1719	6	800 (30%)
	64	351	2356	18	864 (32%)
	128	321	3616	45	992 (37%)
2^{19}	32	277	2032	6	1440 (53%)
	64	261	2567	18	1504 (55%)
	128	265	3890	45	1632 (60%)

With the proposed optimizations, the compiler achieved an II of 1 in all configurations, meaning that the maximum throughput in items/s is equivalent to the reported fmax. We can see from Table 2, that the throughput of the kernel can reach up to 417 million items/s in the smallest configuration. All configurations achieve throughputs that can be considered optimal for this mid-range FPGA family. The largest configuration, spanning 60% of the available embedded RAM on the FPGA chip, achieved a throughput of 265 million items/s. These significantly high numbers are mainly attributed to the simplicity of the synthesized logic and the proposed memory dependency handling technique that allowed for efficient pipelining.

It should be noted that the synthesis tools may report slightly different results for the same configuration when compiled multiple times. The results in Table 2 are obtained from a single compilation process for each configuration.

7. Evaluation

7.1. Accuracy Validation

As the accuracy of the proposed accelerator will be mainly bounded by the amount of on-chip memory allocated for the round table, it is important to validate that the available embedded memory in a typical FPGA chip is sufficient for achieving good accuracy when processing practical data streams. The same Arria 10 GX 1150 FPGA is used as a baseline for accuracy analysis. The amount of embedded RAM available on this chip can be considered in the mid-range of modern FPGA devices. From Table 2, the kernel configurations with the 32-bit key size are selected for accuracy analysis. These configurations are simulated using both synthetic and real datasets. After each simulation run, the heavy hitter summary is sorted to extract the top-1000 items, which are compared to exact results. Two metrics are used in the analysis, as defined below:

Accuracy: The number of correctly identified top-k items divided by k.
Avg. Count Error: The average of relative count errors calculated in all reported items.

Figure 9 reports the accuracy and average error for synthetic data streams. The datasets were generated as Zipfian distributions [41]. The size of all datasets was fixed to 10^7 items, and the Zipfian parameter (α) was varied from 1.0 to 1.6 in intervals of 0.2. The range of α was selected to cover typical values of data skew in several types of real data streams [42]. There are several factors that would influence the best choice of round size r [8]. For simplicity, r was fixed to M in all simulation runs (M—size of the round table in the kernel).

From Figure 9 we can see that all kernel configurations achieved near-perfect accuracy. The worst-case accuracy exceeded 98%, and the worst-case average count error was below 4%.

To further verify the kernel's performance, four different real datasets were also used for accuracy analysis. The properties of the datasets are summarized in Table 3. All of these datasets are easily available and widely used in the analysis of itemset mining algorithms. *Retail* contains market basket transactions data from an anonymous Belgian retail store [43]. *Kosarak* is a collection of click-stream data from a Hungarian online news portal [44]. *Chainstore* contains customer transactions from a major grocery store in California, USA [45]. Finally, *BMS2* contains click-stream data from an anonymous webstore [45]. All of these datasets were originally structured as transactional datasets consisting of several separated transactions, each containing a number of integer items. For the purpose of stream item counting, these datasets were pre-processed to merge the transactions into a serial stream of items. Figure 10 reports the accuracy and average error count results when processing these real data streams.

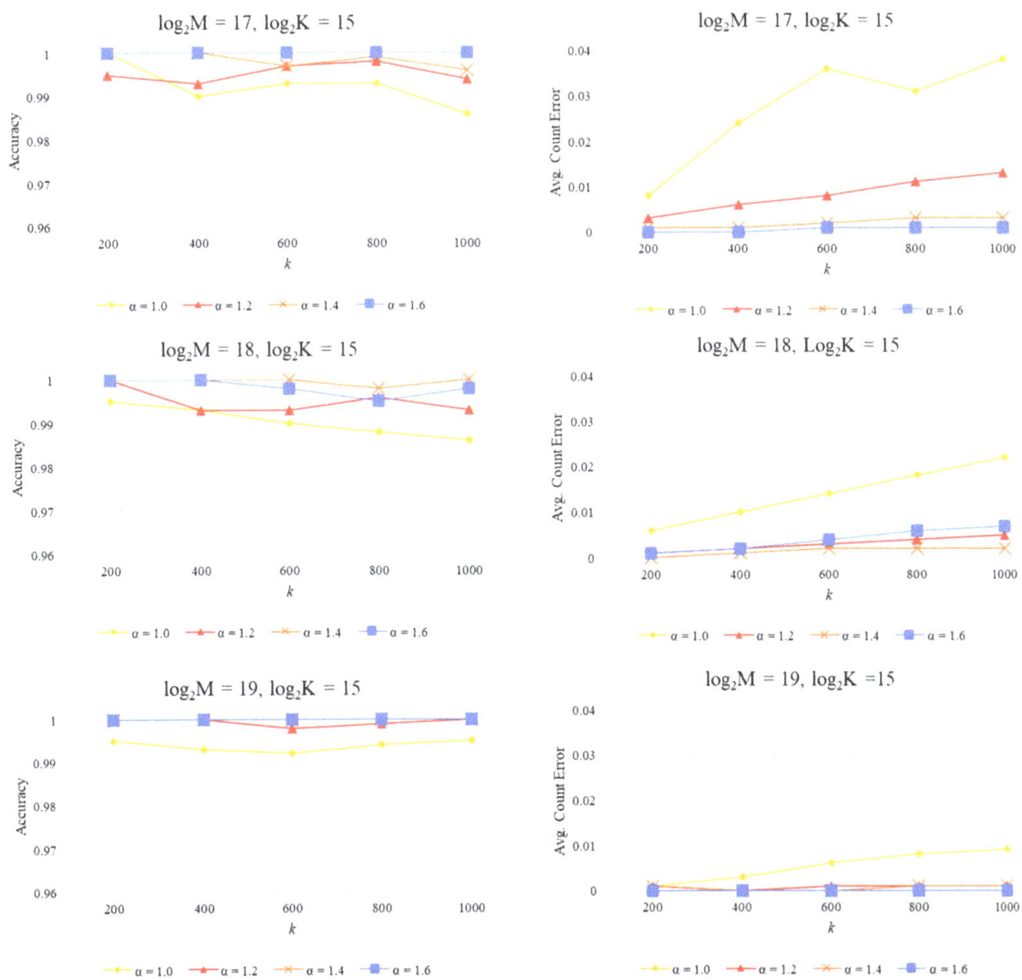

Figure 9. Results for synthetic data streams modelled as Zipfian distributions.

Table 3. Real datasets.

Dataset	Distinct Items	Size
Retail	16,469	908,399
Kosarak	41,270	8,019,015
Chainstore	46,086	8,042,879
BMS2	3340	358,278

We can see from Figure 10 that the kerel maintained a high level of accuracy when processing the real datasets. The worst-case accuracy exceeded 95%, and the worst-case average error count was 2%.

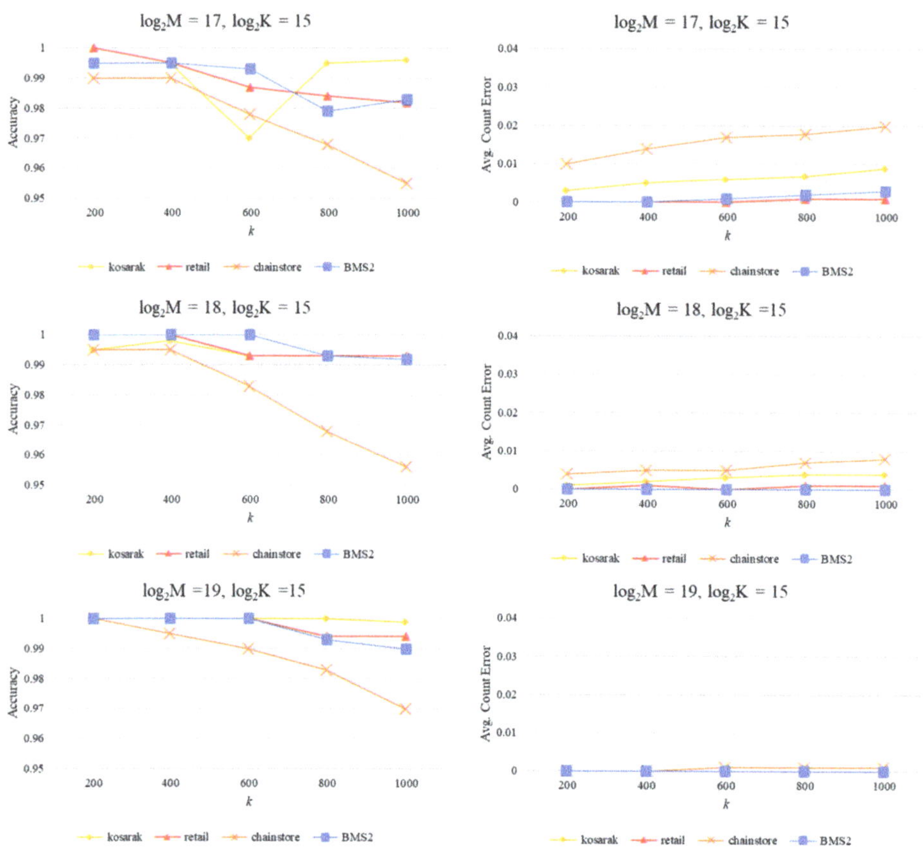

Figure 10. Results for real datasets.

7.2. Comparison with the State-of-the-Art

This section compares the proposed accelerator kernel to the fastest FPGA accelerators previously proposed in the academic literature (see Table 4). As discussed in Section 2, there are a limited number of FPGA accelerators specifically designed for the top-k query problem in data streams. Therefore, we extend the comparison to include generic data stream sketches. These sketches are simple and do not monitor any item key as they do not include a priority queue data structure. Four different metrics are used in the comparison:

Chip utilization: The highest chip utilization percentage of any resource type in the accelerator implementation (mainly logic resources for systolic array implementations and embedded memory for sketch implementations).

Monitored items: The number of item keys that are monitored by the accelerator. In Table 4, the relevant entries are labelled with "none" for generic sketch accelerators without a priority queue.

Key size: The size of the item key in bits. Smaller key sizes will limit the applications of the accelerator. For example, monitoring IPv6 addresses requires 128-bit item keys.

Throughput: The accelerator processing speed is measured in million items/s. It should be noted that, in some of the relevant publications of the accelerators (Table 4), the throughputs were reported in bits/s and calculated by multiplying the update rate by the item key size or by the network packet size in sketches targeting networking applications.

We use the largest kernel configuration from Table 1 for comparison ($M = 2^{19}$ $K = 2^{15}$). Several accelerators in Table 4 are implemented using high-end AMD/Xilinx UltraScale and UltraScale+ FPGA devices. Since our work was based on Intel FPGA technology, the proposed accelerator was re-implemented on a Stratix 10 FPGA for better comparison with previous implementations on high-end FPGAs. When compiling for a Stratix 10 FPGA as a target, the target fmax was set to 600 MHz, and the safe dependence distance m was set to 16.

Table 4. Comparison with published work.

Implementation	Chip Utilization (%)	Monitored Items	Key Size (bits)	Throughput (M Items/s)
Arria 10 (Proposed)	60	32,768	128	265
Stratix 10 (Proposed)	6	32,768	128	542
Arria 10 [10]	51	300	32	276
Arria 10 [11]	40	1200	32	174
Virtex UltraScale+ [18]	8	none	96	415
Stratix 10 [21]	11	none	32	503
Virtex UltraScale [23]	16	none	128	456
UltraScale+ MPSoC [29]	24	2400	32	354

We can see from Table 4 that the proposed accelerator is far more efficient when compared to accelerators that support the top-k query [10,11,29], as it allows one to monitor a significantly larger number of items with larger keys. In fact, the only other accelerator to support 128-bit keys is the accelerator in [23], which is slower than our proposed accelerator and does not support the top-k item query. When implemented on a Stratix 10 FPGA, the proposed accelerator has a 25% higher throughput compared to the average throughout of competing accelerators implemented on high-end FPGAs. The proposed accelerator is also 8% faster than the fastest competing accelerator.

The presented accelerator was designed using an HLS design flow, which usually leads to performance penalties in favor of better design productivity compared to Register Transfer Level (RTL) design flows. Although most of the accelerators in Table 4 were designed and optimized using RTL, the presented accelerator outperformed all of the other accelerators. The fact that the proposed accelerator outperformed existing FPGA accelerators was mainly attributed to the simplicity of the synthesized hardware. Introducing careful modifications to the implemented algorithm facilitated the resolution of several design complexities. While, in a previous section, we presented an empirical analysis based on synthetic and real datasets to validate the accuracy of the optimized algorithm, we note that further mathematical analysis is required to formally define the error bounds and other metrics, such as the time and memory complexity of the optimized algorithm.

8. Conclusions and Future Work

This paper presented the design and implementation of an FPGA HLS accelerator kernel for computing the top-k heavy hitters in data streams. The kernel is based on a novel hardware-optimized algorithm, allowing for an easily achieved pipelined datapath with an initiation interval of 1. The proposed algorithm incorporates several optimizations, such as fingerprinting, optimistic counting, re-hashing, and timestamping to address several hardware-specific complexities that usually limit the performance of data stream item-counting accelerators. In addition, several FPGA-specific design tweaks that resolve memory dependency issues when implementing the accelerator kernel on an FPGA have been presented. These tweaks deploy unique Load-Store Queues (LSQs) that can be easily implemented using HLS code. When synthesized for Intel FPGA devices using Intel HLS compiler and targeting the Arria 10 and Stratix 10 FPGA families, the resulting synthesized hardware was very simple—mainly consuming the embedded memory resources of the FPGA. Hardware synthesis of several configurations of the kernel showed a variety of promising results: First, the high throughput of the kernel was sustained, even for configurations consuming up to 60% of the on-chip memory. Second, the high throughput and low logic footprint were also sustained when scaling the item key size processed by the kernel from 32-bit to 128-bit. Third, accuracy analysis based on synthetic and real datasets showed that near-perfect results are achievable even using the on-chip memory capacities available in mid-density FPGA families. Finally, compared to existing state-of-the-art accelerators, the proposed accelerator is the fastest—with throughput exceeding 540 million items/s. It is also notably superior in terms of features, as it has a larger key size, larger number of monitored heavy hitters, and supports task parallelism.

Future work will first focus on further validation of the proposed algorithm as well as formally defining the error bounds, time, and memory complexity. In addition, we will explore porting the proposed kernel to a cloud application using FPGAs and different types of accelerators, such as Graphics Processing Units (GPUs).

Funding: This research received no external funding.

Data Availability Statement: Not applicable.

Conflicts of Interest: The author declares no conflict of interest.

References

1. Cormode, G.; Hadjieleftheriou, M. Finding frequent items in data streams. *VLDB Endow.* **2008**, *1*, 1530–1541. [CrossRef]
2. Manerikar, N.; Palpanas, T. Frequent items in streaming data: An experimental evaluation of the state-of-the-art. *Data Knowl. Eng.* **2009**, *68*, 415–430. [CrossRef]
3. Muthukrishnan, S. *Data Streams: Algorithms and Applications*; Now Publishers Inc.: Norwell, MA, USA, 2005.
4. Harrison, R.; Cai, Q.; Gupta, A.; Rexford, J. Network-wide heavy hitter detection with commodity switches. In Proceedings of the Symposium on SDN Research, Los Angeles, CA, USA, 28–29 March 2018; pp. 1–7.
5. Liu, B. *Web Data Mining: Exploring Hyperlinks, Contents, and Usage Data*; Springer: Berlin/Heidelberg, Germany, 2011; Volume 1.
6. Shrivastava, N.; Buragohain, C.; Agrawal, D.; Suri, S. Medians and beyond: New aggregation techniques for sensor networks. In Proceedings of the 2nd International Conference on Embedded Networked Sensor Systems, Baltimore, MD, USA, 3–5 November 2004; pp. 239–249.
7. Cormode, G.; Yi, K. *Small Summaries for Big Data*; Cambridge University Press: Cambridge, UK, 2020.
8. Demaine, E.D.; López-Ortiz, A.; Munro, J.I. Frequency estimation of internet packet streams with limited space. In Proceedings of the European Symposium on Algorithms, Rome, Italy, 17–21 September 2002; pp. 348–360.
9. Bustio-Martínez, L.; Cumplido, R.; Letras, M.; Hernández-León, R.; Feregrino-Uribe, C.; Hernández-Palancar, J. FPGA/GPU-based acceleration for frequent itemsets mining: A comprehensive review. *ACM Comput. Surv.* **2021**, *54*, 179. [CrossRef]
10. Ebrahim, A.; Khalifat, J. Fast approximation of the top-k items in data streams using FPGAs. *IET Comput. Digit. Technol.* **2023**, *17*, 60–73. [CrossRef]
11. Ebrahim, A.; Khlaifat, J. An Efficient Hardware Architecture for Finding Frequent Items in Data Streams. In Proceedings of the IEEE International Conference on Computer Design (ICCD), Hartford, CT, USA, 18–21 October 2020; pp. 113–119.
12. Sun, Y.; Wang, Z.; Huang, S.; Wang, L.; Wang, Y.; Luo, R.; Yang, H. Accelerating frequent item counting with FPGA. In Proceedings of the ACM/SIGDA International Symposium on Field-Programmable Gate Arrays, New York, NY, USA, 26–28 February 2014; pp. 109–112.
13. Teubner, J.; Muller, R.; Alonso, G. Frequent item computation on a chip. *IEEE Trans. Knowl. Data Eng.* **2010**, *23*, 1169–1181. [CrossRef]
14. Metwally, A.; Agrawal, D.; El Abbadi, A. Efficient computation of frequent and top-k elements in data streams. In Proceedings of the International Conference on Database Theory, Edinburgh, UK, 5–7 January 2005; pp. 398–412.
15. Sha, M.; Guo, Z.; Wang, K.; Zeng, X. A High-Performance and Accurate FPGA-Based Flow Monitor for 100 Gbps Networks. *Electronics* **2022**, *11*, 1976. [CrossRef]
16. Pontarelli, S.; Reviriego, P.; Maestro, J.A. Parallel d-pipeline: A cuckoo hashing implementation for increased throughput. *IEEE Trans. Comput.* **2015**, *65*, 326–331. [CrossRef]
17. Cormode, G.; Muthukrishnan, S. An improved data stream summary: The count-min sketch and its applications. *J. Algorithms* **2005**, *55*, 58–75. [CrossRef]
18. Sateesan, A.; Vliegen, J.; Scherrer, S.; Hsiao, H.-C.; Perrig, A.; Mentens, N. Speed records in network flow measurement on FPGA. In Proceedings of the 2021 31st International Conference on Field-Programmable Logic and Applications (FPL), Dresden, Germany, 30 August–3 September 2021; pp. 219–224.
19. Chiosa, M.; Preußer, T.B.; Alonso, G. SKT: A One-Pass Multi-Sketch Data Analytics Accelerator. *Proc. VLDB Endow.* **2021**, *14*, 2369–2382. [CrossRef]
20. Tang, M.; Wen, M.; Shen, J.; Zhao, X.; Zhang, C. Towards memory-efficient streaming processing with counter-cascading sketching on FPGA. In Proceedings of the ACM/IEEE Design Automation Conference (DAC), Virtual, 20–24 July 2020; pp. 1–6.
21. Kiefer, M.; Poulakis, I.; Breß, S.; Markl, V. Scotch: Generating fpga-accelerators for sketching at line rate. *Proc. VLDB Endow.* **2020**, *14*, 281–293. [CrossRef]

22. Saavedra, A.; Hernández, C.; Figueroa, M. Heavy-hitter detection using a hardware sketch with the countmin-cu algorithm. In Proceedings of the 2018 21st Euromicro Conference on Digital System Design (DSD), Prague, Czech Republic, 29–31 August 2018; pp. 38–45.
23. Tong, D.; Prasanna, V.K. Sketch acceleration on FPGA and its applications in network anomaly detection. *IEEE Trans. Parallel Distrib. Syst.* **2017**, *29*, 929–942. [CrossRef]
24. Kohutka, L.; Nagy, L.; Stopjaková, V. A novel hardware-accelerated priority queue for real-time systems. In Proceedings of the 2018 21st Euromicro Conference on Digital System Design (DSD), Prague, Czech Republic, 29–31 August 2018; pp. 46–53.
25. Chen, W.; Li, W.; Yu, F. A hybrid pipelined architecture for high performance top-K sorting on FPGA. *IEEE Trans. Circuits Syst. II Express Briefs* **2019**, *67*, 1449–1453. [CrossRef]
26. Yan, D.; Wang, W.-X.; Zuo, L.; Zhang, X.-W. A novel scheme for real-time max/min-set-selection sorters on FPGA. *IEEE Trans. Circuits Syst. II Express Briefs* **2021**, *68*, 2665–2669. [CrossRef]
27. Zazo, J.F.; Lopez-Buedo, S.; Ruiz, M.; Sutter, G. A single-fpga architecture for detecting heavy hitters in 100 gbit/s ethernet links. In Proceedings of the International Conference on ReConFigurable Computing and FPGAs (ReConFig), Cancun, Mexico, 4–6 December 2017; pp. 1–6.
28. Soto, J.E.; Ubisse, P.; Fernández, Y.; Hernández, C.; Figueroa, M. A high-throughput hardware accelerator for network entropy estimation using sketches. *IEEE Access* **2021**, *9*, 85823–85838. [CrossRef]
29. Soto, J.E.; Ubisse, P.; Hernández, C.; Figueroa, M. A hardware accelerator for entropy estimation using the top-k most frequent elements. In Proceedings of the Euromicro Conference on Digital System Design (DSD), Kranj, Slovenia, 26–28 August 2020; pp. 141–148.
30. Gou, X.; Zhang, Y.; Hu, Z.; He, L.; Wang, K.; Liu, X.; Yang, T.; Wang, Y.; Cui, B. A sketch framework for approximate data stream processing in sliding Windows. *IEEE Trans. Knowl. Data Eng.* **2022**, *35*, 4411–4424. [CrossRef]
31. Yang, T.; Zhang, H.; Li, J.; Gong, J.; Uhlig, S.; Chen, S.; Li, X. HeavyKeeper: An accurate algorithm for finding Top-k elephant flows. *IEEE/ACM Trans. Netw.* **2019**, *27*, 1845–1858. [CrossRef]
32. Pagh, R.; Rodler, F.F. Cuckoo hashing. *J. Algorithms* **2004**, *51*, 122–144. [CrossRef]
33. Cho, J.M.; Choi, K. An FPGA implementation of high-throughput key-value store using Bloom filter. In Proceedings of the Technical Papers of 2014 International Symposium on VLSI Design, Automation and Test, Taiwan, China, 28–30 April 2014; pp. 1–4.
34. Chen, X.; Tan, H.; Chen, Y.; He, B.; Wong, W.-F.; Chen, D. Skew-oblivious data routing for data intensive applications on FPGAs with HLS. In Proceedings of the 2021 58th ACM/IEEE Design Automation Conference (DAC), San Francisco, CA, USA, 5–9 December 2021; pp. 937–942.
35. Kiefer, M.; Poulakis, I.; Zacharatou, E.T.; Markl, V. Optimistic Data Parallelism for FPGA-Accelerated Sketching. *Proc. VLDB Endow.* **2023**, *16*, 1113–1125. [CrossRef]
36. Intel®High Level Synthesis Compiler Pro Edition: User Guide. Available online: https://www.intel.com/content/www/us/en/docs/programmable/683456/21-4/pro-edition-user-guide.html (accessed on 1 March 2023).
37. Ebrahim, A. High-Level Design Optimizations for Implementing Data Stream Sketch Frequency Estimators on FPGAs. *Electronics* **2022**, *11*, 2399. [CrossRef]
38. Preußer, T.B.; Chiosa, M.; Weiss, A.; Alonso, G. Using DSP Slices as Content-Addressable Update Queues. In Proceedings of the 2020 30th International Conference on Field-Programmable Logic and Applications (FPL), Gothenburg, Sweden, 31 August–4 September 2020; pp. 121–126.
39. Huang, H.; Wang, Z.; Zhang, J.; He, Z.; Wu, C.; Xiao, J.; Alonso, G. Shuhai: A Tool for Benchmarking High Bandwidth Memory on FPGAs. *IEEE Trans. Comput.* **2021**, *71*, 1133–1144. [CrossRef]
40. Dietzfelbinger, M.; Hagerup, T.; Katajainen, J.; Penttonen, M. A reliable randomized algorithm for the closest-pair problem. *J. Algorithms* **1997**, *25*, 19–51. [CrossRef]
41. Zipf, G.K. *Human Behavior and the Principle of Least Effort*; Martino Fine Books: Eastford, CT, USA, 1949.
42. Cormode, G.; Muthukrishnan, S. Summarizing and mining skewed data streams. In Proceedings of the the 2005 SIAM International Conference on Data Mining, Newport Beach, CA, USA, 21–23 April 2005; pp. 44–55.
43. Brijs, T.; Swinnen, G.; Vanhoof, K.; Wets, G. Using association rules for product assortment decisions: A case study. In Proceedings of the ACM SIGKDD International Conference on Knowledge Discovery and Data Mining, San Diego, CA, USA, 15–18 August 1999; pp. 254–260.
44. Frequent Itemset Mining Dataset Repository, University of Helsinki. Available online: http://fimi.cs.helsinki.fi/data/ (accessed on 2 October 2021).
45. Fournier-Viger, P.; Lin, J.C.-W.; Gomariz, A.; Gueniche, T.; Soltani, A.; Deng, Z.; Lam, H.T. The SPMF open-source data mining library version 2. In Proceedings of the Joint European Conference on Machine Learning and Knowledge Discovery in Databases, Riva del Garda, Italy, 19–23 September 2016; pp. 36–40.

Disclaimer/Publisher's Note: The statements, opinions and data contained in all publications are solely those of the individual author(s) and contributor(s) and not of MDPI and/or the editor(s). MDPI and/or the editor(s) disclaim responsibility for any injury to people or property resulting from any ideas, methods, instructions or products referred to in the content.

electronics

MDPI

Article

Accurate Multi-Channel QCM Sensor Measurement Enabled by FPGA-Based Embedded System Using GPS

Adrien Bourennane [1,*], Camel Tanougast [1,*], Camille Diou [1] and Jean Gorse [2]

[1] LCOMS, Université de Lorraine, 57000 Metz, France
[2] Pesage Lorrain Continu et Discontinu, 57070 Saint-Julien-Lès-Metz, France
* Correspondence: adrien.bourennane@univ-lorraine.fr (A.B.); camel.tanougast@univ-lorraine.fr (C.T.)

Abstract: This paper presents a design and implementation proposal for a real-time frequency measurement system for high-precision, multi-channel quartz crystal microbalance (QCM) sensors using a field programmable gate array (FPGA). The key contribution of this work lies in the integration of a frequency measurement and mass resolution computation based on Global Positioning System (GPS) signals within a single FPGA chip, utilizing Input/Output Blocks to incorporate logic QCM oscillator circuits. The FPGA design enables parallel processing, ensuring accurate measurements, faster calculations, and reduced hardware complexity by minimizing the need for external components. As a result, a cost-effective and accurate multi-channel sensor system is developed, serving as a reconfigurable standalone measurement platform with communication capabilities. The system is implemented and tested using the FPGA Xilinx Virtex-6, along with multiple QCM sensors. The implementation on a Xilinx XC6VLX240T FPGA achieves a maximum frequency of 324 MHz and consumes a dynamic power of 120 mW. Notably, the design utilizes a modest number of resources, requiring only 188 slices, 733 flip-flops, and 13 IOBs to perform a double-channel sensor microbalance. The proposed system meets the precision measurement requirements for QCM sensor applications, exhibiting low measurement error when monitoring QCM frequencies ranging from 1 to 50 MHz, with an accuracy of 0.2 ppm and less than 0.1 Hz.

Keywords: embedded systems; FPGA-based applications; IOB interfaces; quartz crystal microbalance; frequency measurement; GPS

Citation: Bourennane, A.; Tanougast, C.; Diou, C.; Gorse, J. Accurate Multi-Channel QCM Sensor Measurement Enabled by FPGA-Based Embedded System Using GPS. *Electronics* **2023**, *12*, 2666. https://doi.org/10.3390/electronics12122666

Academic Editors: Andres Upegui, Andrea Guerrieri and Laurent Gantel

Received: 30 April 2023
Revised: 11 June 2023
Accepted: 12 June 2023
Published: 14 June 2023

Copyright: © 2023 by the authors. Licensee MDPI, Basel, Switzerland. This article is an open access article distributed under the terms and conditions of the Creative Commons Attribution (CC BY) license (https://creativecommons.org/licenses/by/4.0/).

1. Introduction

Nowadays, many industrial applications that involve the handling or processing of physical, chemical, or biological substances rely heavily on high-precision measurement instrumentation. One of the most commonly used systems for this purpose is the QCM (quartz crystal microbalance), which detects the resonance and frequency shifts of quartz crystal resonator (QCR) sensors. In this context, QCM sensors are widely used in different application domains. QCM sensors find wide application in various domains due to their high sensitivity and real-time capability of measuring minute mass changes (typically in the order of a few ng/cm^2) within a broad dynamic range (100 µg/cm^2). This makes them particularly attractive for applications such as bio-sensors, analysis of biomolecular interactions, and studying cell–substrate interactions [1]. Usually, to perform high-precision measurements, accurate frequency (/time) measurement techniques are employed using electronic resonators based on circuits containing capacitors, resistors, and/or inductors [1]. These circuits generate alternating current by periodically fluctuating between two voltage levels. Oscillators working with optimal stability rely on vibrating quartz crystals, which exhibit a stable frequency when a direct current is applied. Similarly, a piezoelectric oscillator circuit uses a piezoelectric crystal in combination with electronic passive components to generate a stable frequency depending on crystal properties and environmental conditions [2]. Factors such as temperature, pressure, acceleration, radiation, electric fields, and electromagnetic fields can introduce variations in the nominal generated frequency oscillation. As a result, sensors based on piezoelectric oscillators offer accurate measurement of these physical variables [3]. Therefore, piezoelectricity based on the quartz crystal microbalance is one of the most popular mass sensing techniques in industrial applications, including gas and liquid sensors [2–4] and electronic tongues [5]. These applications include molecular recognition [6–8] and food quality control [4,9–11]. QCM is a low-cost and highly sensitive mass measurement technique that was discovered in 1959 by Sauerbrey [12]. Sauerbrey established a relationship between the mass on the surface of the crystal and its resonance frequency. More precisely, as depicted in Figure 1, the addition of mass distributed over the quartz crystal surface alters the nominal oscillation frequency. This frequency variation can be described by the following Sauerbrey Equation (1):

$$\Delta f = \frac{-2 \times {f_0}^2}{A \times \sqrt{\rho_q \times \mu_q}} \times \Delta m. \quad (1)$$

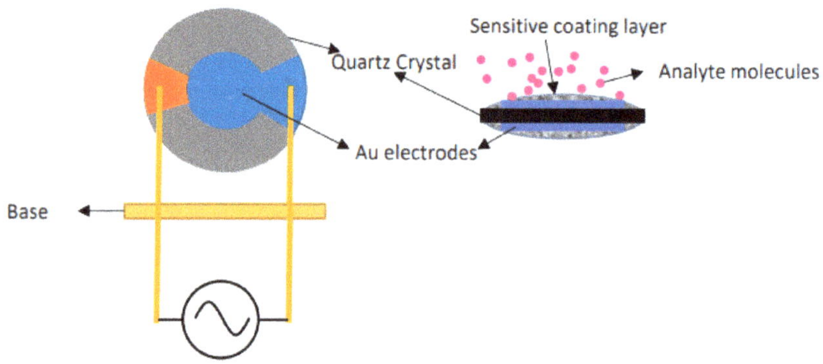

Figure 1. Basic working principle of quartz crystal microbalance sensor [13].

Here, Δf represents the normalized frequency change (Hz) as a function of the mass change Δm (gram), f_0 is the resonant frequency (Hz), A is the piezoelectrically active crystal area (area between electrodes, cm^2), ρ_q is the density of quartz (2.648 g/cm^3), and μ_q is the shear modulus of quartz for AT-cut crystal (2.947 $\times$ 10^{11} g·cm^{-1}·s^{-2}).

Therefore, the resonance and subsequent frequency shift of the quartz crystal resonator is detected by a QCM measurement system. A QCM crystal consists of a thin quartz crystal with metallic electrodes of a certain thickness on both sides. This pellet is produced with different thicknesses, resulting in different frequencies. Gold is often used for the electrodes due to its resistance to corrosive environments [14]. There are three main electronic techniques used for frequency shift measurements: impedance measurement, quartz crystal microbalance with dissipation (QCM-D), and oscillator-based measurements [15]. Among these techniques, impedance measurement provides the most precise results for resonance frequency analysis [16]. It involves applying a sweeping frequency signal to a quartz crystal resonator and collecting impedance spectrum (or admittance) data to determine the resonant frequency and dissipation outputs. QCM-D is a type of quartz crystal microbalance based on the ring-down technique. It is often used to determine film thickness in a liquid environment, such as the thickness of an adsorbed protein layer. It can be used to study other properties of the sample, such as its softness. The QCM-D technique allows measurement of several times per second in a vacuum, gaseous, or liquid environment [17]. Additionally, it is possible to switch between fundamental frequency and overtones [18]. Although QCM-D and impedance measurement systems are efficient and commercially available, they are often expensive and cumbersome. They are not adequate for on-site use. The principles of oscillator-based measurement, distinguishing inverting and non-inverting amplifier oscillators, are illustrated in Figure 2. On the one hand, the inverting amplifier, known as a Pierce oscillator (shown in Figure 2a), adds a 180° phase shift which is compensated by the feedback network (based on R_a, C_1, C_2 passive components and the quartz crystal) to meet the phase requirement in the Barkhausen criterion. On the other hand, the non-inverting amplifier (shown in Figure 2b) acts on the sensor as a series resonator satisfying the phase condition at the series resonance frequency by only using resistor components (R_a, R_b). Figure 2c illustrates another non-inverting amplifier known as a Colpitts oscillator, where the sensor functions as a high-quality inductor through its connection in parallel with R_1, C_2 passive elements.

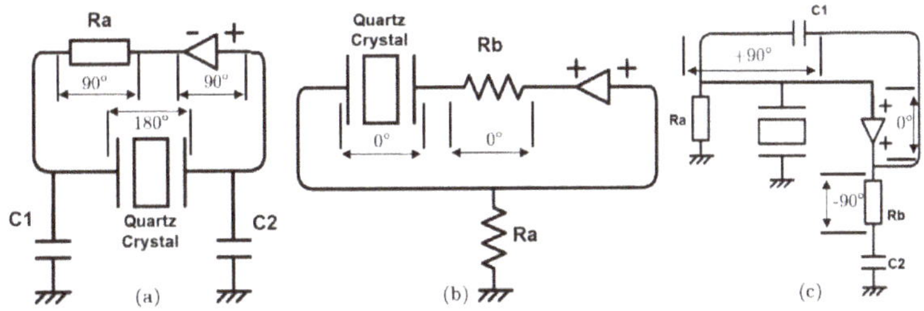

Figure 2. Typical oscillator circuits: (**a**) with an inverting amplifier (Pierce oscillator); (**b**) with a non-inverting amplifier, and (**c**) with a Colpitts oscillator [16,19].

The QCM is widely used due to its extreme sensitivity to the characteristics of the materials it comes into contact with, leading to shifts in its resonant frequency. However, the effectiveness of the QCM is constrained by the noise specifications of the crystal oscillator and the resolution of the frequency counter employed to measure frequency variations. Usually, the standard QCM System is a stand-alone instrument with the built-in quartz crystal oscillator electronics, frequency counter, and CPU/microcontroller ensuring the measurement, the monitoring, and the display (on a front panel) of the shifts in resonance frequency, which is dependent on the material with which the QCM is in contact. Consequently, an input stimulus induces a frequency shift in the sensor. Therefore, precise quantification of changes in the input stimulus is achievable, provided an appropriate frequency counter/meter is utilized. Unfortunately, it is well known in the field of time–frequency metrology that attaining higher measurement accuracy necessitates longer measurement times. To mitigate this, QCM systems incorporate a phase-locked loop (PLL) electronic circuit, which reduces the measurement time [20]. Nevertheless, such systems are neither cost-effective nor suitable for developing a multi-channel QCM system. Each QCM would require a quartz crystal resonator oscillator, a PLL, a low-pass filter, and an amplifier circuit.

Static random-access-memory-based field programmable gate array (SRAM-Based FPGA) technology provides a parallel computation capability which offers performance improvements while ensuring flexibility compared with traditional CPU processing architectures. Moreover, FPGAs provide Input/Output Blocks (IOBs) which can be used to implement additional logic with CLBs to improve design performance by increasing available logic and routing resources. Previous works show interest in using FPGA for the integration of a frequency measurement technique providing a trade-off between performance and accurate measurement [21]. Similarly, several works have explored the utilization of FPGAs in QCM systems for conventional counter-based frequency measurement, with or without compensation circuits [22,23]. For example, a low-cost prototype of a multi-channel quartz crystal microbalance data acquisition system for QCM sensor investigation was developed in 2018 [10]. It uses a totally external oscillator to keep the oscillation down. The 16-bit time counters of the PIC16F allow frequency measurement to be performed by QCM sensors with a sensitivity of 1 Hz. However, all of these prior works required additional external chips to realize the QCM oscillators and generate the reference signal based on a subdivision of a highly accurate local clock oscillator.

This paper presents a proposal to implement a commonly used Pierce-gate crystal oscillator based on a quartz crystal using configured Input/Output Blocks (IOBs) within an FPGA. By incorporating digital inverters and a feedback resistor, the inner digital CMOS inverter can be linearized, effectively transforming the logic inverter gate into an analog amplifier. This approach allows for the utilization of low-cost external passive components (such as C1 and C2 reactance and Rs) that satisfy the Barkhausen criteria.

Input/Output Blocks connect internal FPGA architecture to the external design via interfacing pins, eliminating the need for external chip oscillators, such as resonators, PLLs, amplifier circuits, or filters. Moreover, the FPGA's logic elements, which serve as the fundamental building blocks, can be programmed to carry out different functions as required by the design, enabling the implementation of accurate frequency measurement using the GPS as a reference signal. The main novelty of the proposed FPGA-based system lies on the use of internal IOBs and its ability to perform a multi-QCM measurement system composed of several oscillators, each equipped with a QCM. Within this system, Pierce oscillators utilize on-chip inverting amplifiers within the IOBs of the FPGA, striking a balance between achieving accurate frequency measurements (which can be enhanced through fine measurements based on a ring oscillator or time-to-digital approach) while minimizing the use of logic resources within the FPGA and external components. Therefore, the frequency measurement is accomplished through the implementation of a 32-bit reconfigurable reciprocal frequency meter architecture, which relies on the GPS reference signal and the measurements taken when it is connected to different QCMs.

The remaining sections of the paper are structured as follows. Section 2 outlines the proposed system, which integrates parallel quartz crystal oscillators using only IOBs connected to multiple QCMs simultaneously. This section describes the reciprocal counter implemented in FPGA (with the potential for enhanced accuracy through the implementation of a time-to-digital converter (TDC) in the FPGA). Section 3 investigates the resonant conditions for various QCMs, providing details on the experimental setup, measurement results, and subsequent discussion of the findings. Finally, Section 4 presents the conclusion of the study along with directions for future work.

2. Frequency Measurement Electronic System

Figure 3 illustrates the overall electronic measurement system of the proposed multi-channel QCM data acquisition system, which incorporates two QCM resonators. The embedded frequency measurement system comprises both hardware and software components.

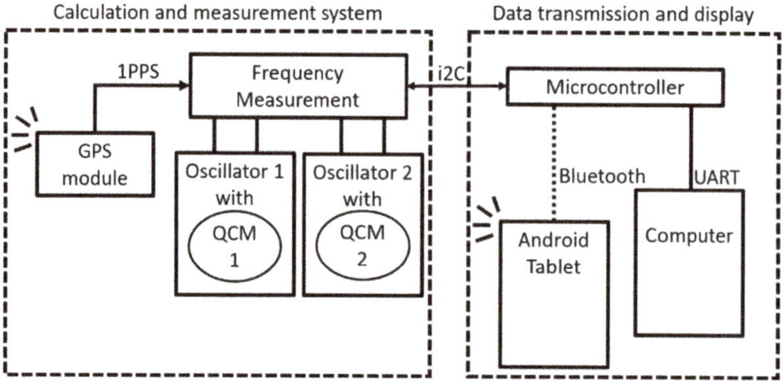

Figure 3. Block diagram of multi-channel QCM data acquisition system.

The embedded hardware employed in this study utilizes an accurate FPGA-based frequency measurement system implemented on the Xilinx Virtex 6 FPGA ML605 stand-alone platform [24]. The software part, executed on a microcontroller, is responsible for collecting frequency data and transmitting this to a display. This proposed electronic communication system incorporates an embedded processor that utilizes Bluetooth communication with a custom-developed Android application. This application receives and stores data from the frequency measurement hardware system in the form of a spreadsheet. This electronic system ensures the transmission of data acquired from the sigma-delta converter through the utilization of the I2c protocol. More precisely, serial data (SDA) and clock (SCL) signals are used to perform the I2c protocol ensuring data transmissions. One on-board microcontroller on the red PIC32MX470 development board manages the communication with the FPGA via the I2c protocol. The timer configuration is used to set the measuring rate. The UART link is used to send the results from the microcontroller to the computer in order to fill in a spreadsheet. The timer defines the delay between each acquisition request for digital frequency values. The Android application manages the reception of measured values via the Bluetooth protocol and displays in one tablet device. In summary, the microcontroller is responsible for managing frequency measurements, which are conducted by the FPGA directly connected to multiple QCM measurement channels.

A multi-channel reciprocal counter is implemented within the FPGA Virtex-6, utilizing a 200 MHz local clock reference signal. The timegate (measurement time) for counting rising edges of the reference signal is set to one second, corresponding to a 1PPS GPS signal received with a jitter of approximately 20 ns from the MediaTek GPS Chipset MT3339 of the Adafruit GPS module [25]. This received GPS reference signal provides an efficient, stable, and cost-effective solution for generating an accurate timegate reference signal, enabling high-frequency resolution in frequency measurements. Figure 4 showcases a photo of the proposed embedded hardware system for QCR oscillator multi-channel frequency measurements.

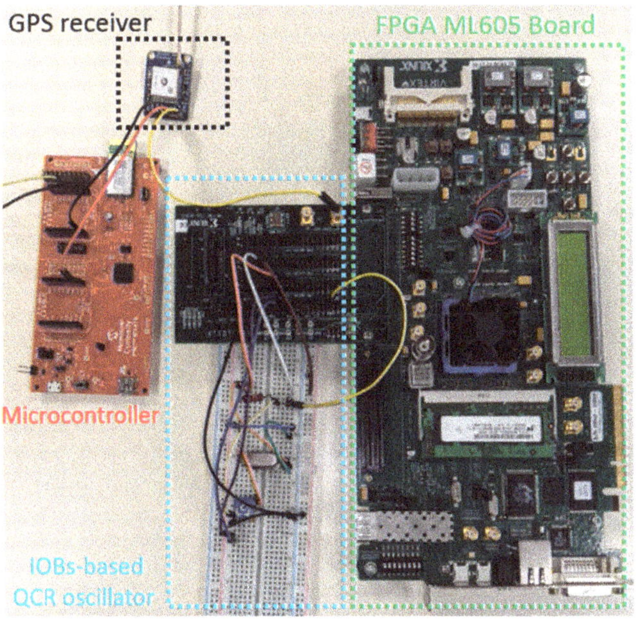

Figure 4. Photo of the proposed resonator multi-channel reciprocal counter implemented in the FPGA-Virtex-6-based GPS clock reference signal.

2.1. Frequency Measurement System

Usually, low-cost FPGA-based frequency measurement relies on a frequency counter that incorporates timers and logical counters. The basic digital measurement of frequency involves counting the number of rising edges of the input signal during a predetermined time interval, utilizing a stable clock reference signal. The resonance frequency can then be obtained using Equation (2), where f represents the measured resonance frequency, N denotes the measured number of input pulses, and t indicates the measurement time.

$$f = \frac{N}{t}. \quad (2)$$

Depending on the frequency of the clock reference signal, the accuracy of the measurement improves as the duration of the measurement time increases. If we use a basic frequency counter (according to Equation (2)) with a measurement window of one second, we will reach a maximum accuracy of ±1 Hz. However, for QCM applications that require higher frequency resolution and/or shorter measurement time, modern frequency counters employ the reciprocal counting method [26]. Unlike previous approaches that solely rely on a high-frequency reference signal, the proposed reciprocal counter utilizes two signal references: a high-frequency clock signal for frequency calculation and the received GPS signal for period measurement. This approach offers an alternative solution for achieving more accurate measurements without the need for subdivision of the high clock reference signal. Figure 5 illustrates the reciprocal counting method using the GPS Signal.

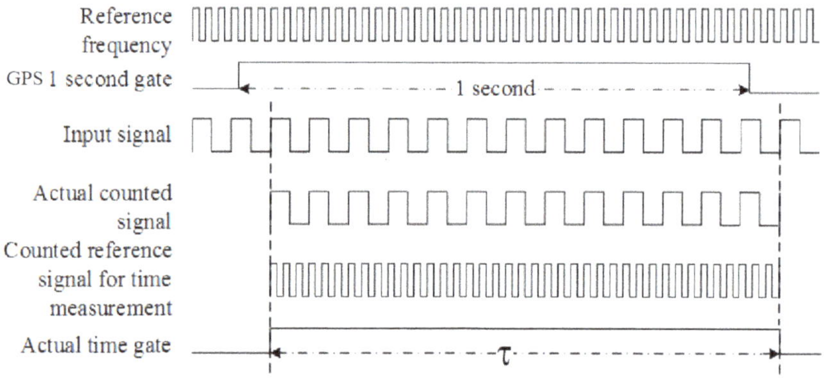

Figure 5. Reciprocal counter based on an input reference GPS signal.

For a duration of one second, each clock edge of the input signal is counted N times. Simultaneously, a reference signal is also counted for the exact same duration. This duration, referred to as τ, represents the time delay between the first rising edge of the input signal following the rising edge of the one-second period signal, and the first rising edge of the input signal following the falling edge of the 1 Hz signal. The main advantage of this method is its resolution, which remains unaffected by the input frequency and can be enhanced through the utilization of low-cost FPGA-based digital counting techniques (DCTs) for time stamping the start and stop edges of the input signal. Moreover, the error remains constant across the entire range of input signal frequencies and can be reduced as the reference clock frequency increases or as the gate time extends.

Figure 6 illustrates the internal architecture of the FPGA reciprocal counter designed for measuring the frequency of a single QCR oscillator. This design incorporates three digital counters, each 32 bits in size, responsible for counting the various edge events of the input oscillation signal. These counters enable the calculation of the frequency value or frequency shifts resulting from the input QCM stimulations connected to the FPGA's IOBs. The first counter, denoted ValeurFrequenceRef, provides the count result of the high frequency reference delineated by the input GPS signal. This measurement provides a real-time accurate measurement of the reference frequency. The second counter, referred to as C_Freq, counts the number of rising edges of the signal to be measured (the output oscillation signal for one Pierce QCM oscillator) within the time period τ. Finally, the last counter performs the task of counting the number of rising edges of the reference signal during the designated period τ in order to obtain the frequency measurement of the input signal by considering Equation (2).

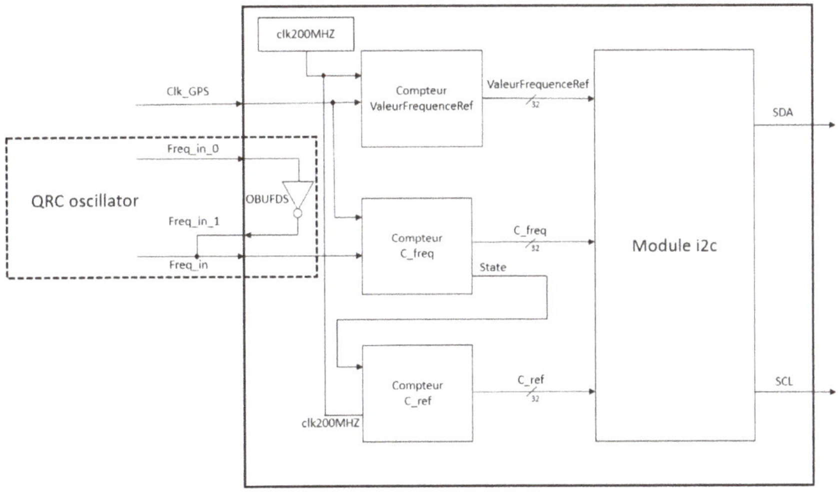

Figure 6. Architecture of reconfigurable frequency meter.

Behavioral simulations and timing were conducted to evaluate the accuracy of the proposed test using the Xilinx Virtex-6 ML 605 platform. These simulations demonstrated the highly accurate frequency measurement of the oscillating QCM signal obtained from an on-chip FPGA IOB logic inverter. The simulation results of the proposed architecture, which utilizes the GPS-based reciprocal counting method, are presented in Figure 7. These results were obtained through the utilization of VHDL description and the Xilinx FPGA ISE Environment design tool.

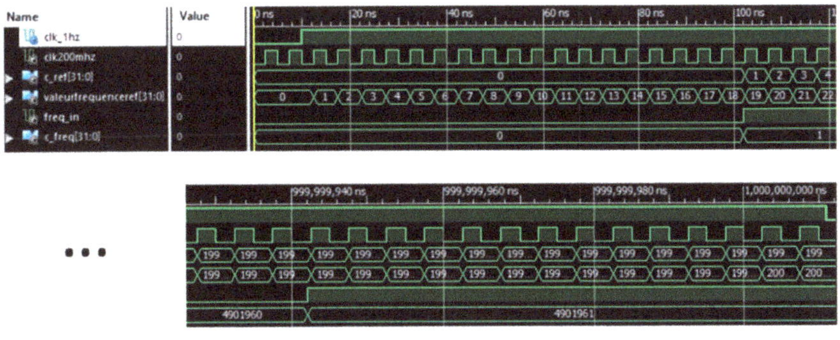

Figure 7. Timing behavioral simulation results of accurate frequency measurement based on reciprocal counter design.

The design of the proposed reconfigurable frequency meter incorporates the GPS and oscillating QCM signals as inputs (represented by clk_1 hz and freq_in signals in Figure 7). The behavioral simulation results demonstrate the functionality and the accurate values obtained by three counters (C_Ref, ValeurFrequenceRef, C_Freq) in accurately measuring the frequency of the freq_in signal. These measurements take advantage of the precision of the GPS sensor (clk_1 hz signal) and a local clock frequency of 200 MHz (clk200mhz signal).

The accurate frequency calculation is performed according to Equation (3), as depicted in Figure 5.

$$f = \frac{C\_Freq * ValeurFrequenceRef}{C\_Ref}. \quad (3)$$

2.2. Oscillator Realization

The resonance time of QCM crystals exhibits a low oscillation amplitude, typically measured in millivolts. Therefore, sensors are connected directly to the oscillator circuit to preserve the signal level and shape. However, the crystal signal can suffer from attenuation and interference as it traverses various contact points such as the crystal holder and connectors before reaching the circuit. To address this issue, great care has been taken in selecting the connecting components, and proactive measures have been employed to position the oscillator components away from strong magnetic fields and power connections.

As mentioned, the on-chip IOB inverters of the FPGA are configured as OBUFDS elements [27] to directly facilitate the implementation of Pierce QCR oscillators. This novel approach, not previously proposed in the existing literature on multi-channel QCM measurement systems, enables the realization of low-cost QCR oscillators. Figure 8 describes the proposed IOBs' configuration and connection to create multi-channel QCR oscillators. Pierce oscillators are implemented with QCM connected to the FPGA realizing the inverting gate for the oscillator circuits responsible for generating the frequency oscillation signals to be measured. For this purpose, some input pins of the FPGA are connected to QCR by using the connection pins of the FMC XM105 expansion card (see Figure 4). For the input frequency signal to be measured (i.e., the oscillator output signal), "Clock Capable" pins ending in "_CC_" are used. For the OB output signal of the OBUFDS component, pins ending in "_N" must be used, and for the O output signal of the OBUFDS component, pins ending in "_P" should be used. In order to ensure adjacent output ports, the positive and negative outputs are positioned on the same IOB side.

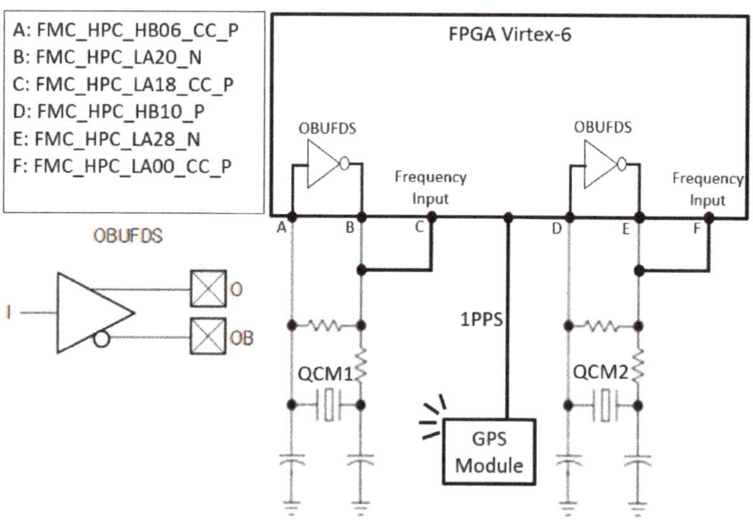

Figure 8. Design of Pierce-QCR-oscillator-based inner logic inverters of IOB FPGA.

3. Implementation and Experimental Results

The proposed reconfigurable standalone measurement platform with communication capabilities combines both software resources, which bring the necessary versatility, and the logic hardware resources in which the accurate frequency measurement unit is implemented. To implement and test with multiple QCM sensors, the proposed design based on the Xilinx XC-6 ML 605 technology is integrated as a new core with specific components as described by the corresponding block diagram shown in Figure 9. The proposed block design includes the following four main modules:

- The Frequencemetre module, which corresponds to the proposed reciprocal counter and delivers a 32-bit sequence of three counter values representing the digital frequency measurement, as determined by Equation (3).
- The I2c_Slave_top module, which represents a hierarchical communication using the I2c protocol that is connected to and exchanges data with the microcontroller for the purpose of frequency measurement display. This module receives the counter values from the Frequencemetre module and transmits them via the I2c communication link to the microcontroller (PIC32MX) for further processing and display of the corresponding normalized frequency change (in Hz) based on the QCM microbalance oscillation.
- The OBUFDS block, which is the inner logic oscillator circuit that provides the oscillating QCM signal while reducing the need for extra external logic circuits to function as an inverting amplifier oscillator.

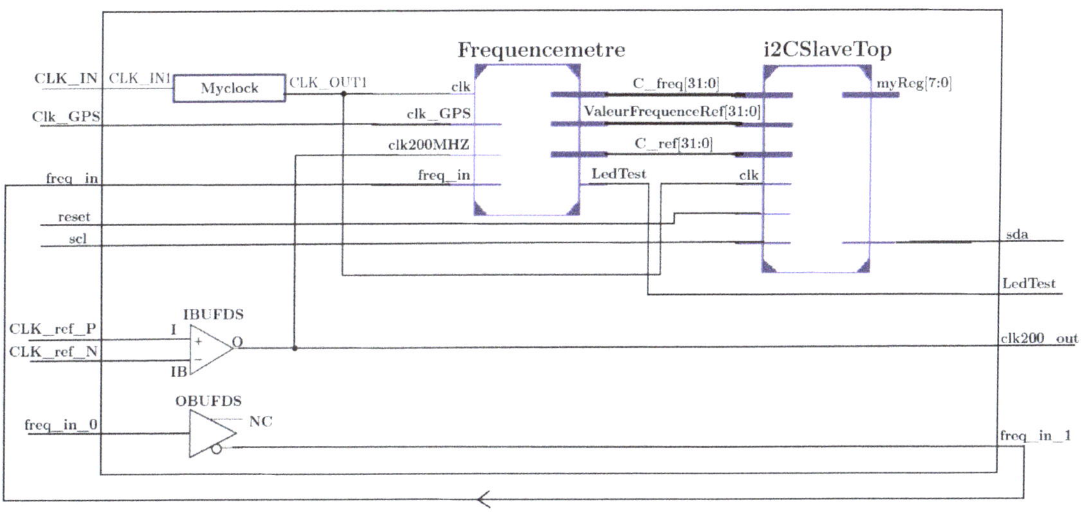

Figure 9. Design block diagram of GPS-based frequency measurement by QCM sensors.

As shown in Figure 9, the configured IOBs function as external oscillators ensuring multi-channel measurement as illustrated in Figure 8. For this purpose, two IOBs are necessary to achieve a single-channel QCM oscillator. More precisely, one IOB (OBUFDS block) is configured as one logic inverter, which is combined with the external passive elements (C1 and R1 as described in Figure 2) to create an on-chip piezoelectric QCM oscillator within the FPGA. The second IOB is configured as a single-input buffer (IBUFDS-configured IOB), serving as an intermediary element that ensures the signal source remains unaffected by the load attributes while delivering a voltage and current similar to what it receives at its input.

The proposed accurate multi-channel frequency measurement is described by using hardware description language (VHDL), and the final binary configuration file is implemented on Xilinx Virtex 6 XC6VLX240T-1FFG1156 FPGA. Table 1 provides a breakdown of the required logic synthesis resources for the system, which include two QCR oscillators connected to the FPGA. The resource utilization is as follows: 733 Slice registers, no DSP multipliers, and no block RAM. This results in a low-cost logic consumption, with 188 slices, 733 flip-flops, and 13 IOBs, all operating at a maximum frequency of 324 MHz. The architecture exhibits a dynamic power of 0.120 W, with a total supply power of approximately 3.618 W. Compared with a similar previous work [28], the proposed system for multi-channel measurement eliminates the need for Block RAM or specific digital clock managers (DCMs) associated with multiple GCLKs used as delay-locked loop (DLL) for accurate measurement [28]. Moreover, due to the Xilinx FPGA technology, the proposed system consumes over five times less dynamic power than other DCM- or DLL-based systems which require a minimal dynamic power of 727 mW and 662 mW, respectively, with a 100 MHz clocking frequency [29].

Table 1. Comparison of FPGA Resource Utilization for Two Parallel QCMs.

	FPGA Technology	GCLK	DCMs	BRAMs	LUTs	Slices	IOBs	FFs	Dynamic PWR (mW)	Fmax (MHz)
Proposed Work	Virtex-6 XC6VLX240T	1 (12.5%)	0 (0%)	0 (0%)	527 (1%)	188 (1%)	13 (2%)	733 (1%)	120	324.4
Ref. [22]	Virtex-4 4vlx25fft668-10	2 (6%)	1 (12%)	X	1625 (7.5%)	922 (8.5%)	8 (1.5%)	774 (3.5%)	NC	102.963
Ref. [28]	Spartan-3 XC3S200	NC	1 (25%)	1 (9%)	460 (12%)	230 (12%)	5 (3%)	460 (12%)	NC	200

To validate the system's performance, a test platform consisting of multiple QCM sensors in a homogeneous liquid (distilled water) was utilized. Figure 10 displays the measurement obtained by the proposed multi-channel QCM from output signals of 4 MHz and 10 MHz QCR oscillators. These signals were obtained using OBUFDS differential inverters to perform 4 MHz and 10 MHz on-chip FPGA Pierce oscillator using only external passive components (see Figure 4). We observed that the IOBs of the FPGA ensure the oscillation of several quartz crystals without the need for additional external logic circuits, as commonly utilized in previous QCM measurement systems. Therefore, the proposed integrated QCRs maintain the oscillation of the signals; these are directly measurable by the Frequencemetre module within the FPGA. These tests validate the functionality of the FPGA-based multi-channel QCM measurement system, which operates effectively without external QCRs and provides accurate real-time frequency measurements.

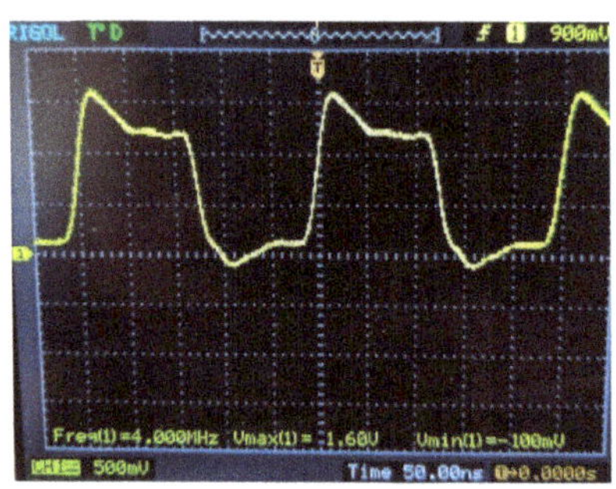

(**a**)

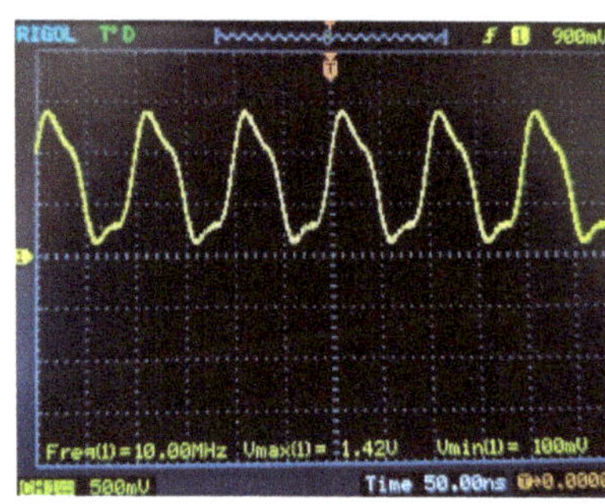

(**b**)

Figure 10. Frequency output signals of the QCR oscillators of the proposed multi-channel QCM measurement: (**a**) 4 MHz QCR and (**b**) 10 MHz QCR.

In order to assess the accuracy of the proposed FPGA-based frequency meter design, we conducted a series of comparative measurements. We used an input signal from a 10 MHz MicroCrystal OCXO oscillator, reference "OCXOV-AV5-10.000", which provided a frequency stability of ±0.2 ppm. This reference signal is measured both by a frequency meter (Rohde and Schwarz Hameg HM8123 (OCXO Version)) and the proposed FPGA Virtex-6 frequency measurement system. The HM 8123 counter frequency measurement error is 1.25×10^{-8}. It represents the frequency instability of the counter, in the temperature range of 0–50 °C. A 120-second series of measurements can be seen in Figure 11. On all the measurements taken, it can be seen that the difference (Δf) between the mean value of measured frequencies by the HM8123 m and the mean values measured by the frequency meter implemented in the FPGA is less than 0.1 Hz. In Figure 11, Δf is approximately 0.09 Hz. We can therefore consider that under normal operating conditions, the accuracy of our system is better than 0.1 Hz.

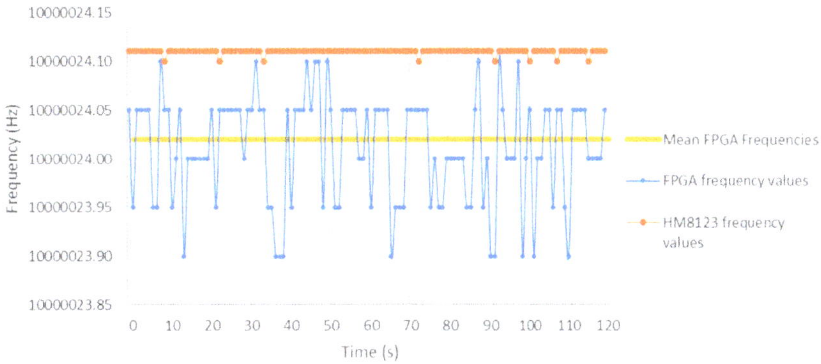

Figure 11. Frequency measurement results for a gate time of one second.

4. Comparison with Similar Works

Several recent systems have been proposed to measure frequency shifts in QCM systems based on QCR [22,23,28,30–34]. Table 2 provides a comparison between the proposed accurate multi-channel QCM system and state-of-the-art works in terms of additional hardware resources, number of channels, number of external logic oscillator circuits (QCRs), reference time base source, accuracy, and required FPGA logic resources. Based on Table 2, it is evident that the proposed system achieves one of the highest accuracies. This is primarily attributed to the utilization of a GPS signal and the incorporation of QCRs specifically designed within the FPGA. These features enable the system to conduct multi-channel frequency measurements using just a single FPGA. In fact, compared with previous works, the achieved accuracy is better than 0.1 Hz, which is accomplished with only a reciprocal counter. Furthermore, compared with similar works, the proposed FPGA design utilizes 79% fewer slices, 5% fewer flip-flops, and 67% fewer LUTs compared with the architecture presented in [22]. Similarly, the proposed FPGA architecture design necessitates 45% fewer slices compared with the system suggested in [28]. Therefore, compared with previous systems, the proposed multi-channel QCM measurement system offers low-cost, highly accurate frequency measurement for multiple interconnected sensing QCM crystals in parallel for sensor investigations. Consequently, the proposed system provides a better trade-off in terms of performance and required logic resources while offering a cost-effective solution as it only requires external passive components to achieve multiple inverting QCM oscillators, allowing multi-channel QCM measurement using a single FPGA device. Moreover, the main advantage of the proposed system is that it enables multi-channel frequency measurement without the need for specific embedded blocks such as DSP, BRAM, DLL, or additional external logic circuits such as oscillators, PLLs, microcontrollers, etc. However, a limitation of the proposed system is its performance when ensuring simultaneous measurement of a large number of QCR frequencies in the FPGA, which requires high parallel computation and a large number of IOB (Input/Output Buffer) clocks.

Table 2. Comparison of QCR-based QCM measurement systems.

References	Hardware	Reference Time Base Source	Active QCR	Techniques	Precision Achieved	Resources for Frequency Measurement	Reconfigurability with Multi-Channel System
[28]	Xilinx Spartan 3	External 50 MHz oscillator	External	Conventional frequency counter, differential delay lines (DLL)	0.05 Hz	230 slices, 1 DCM, 9 IOBs, 17,280 BRAMs	Yes
[20]	Low-pass filter, amplifier, PLL with VCO, microcontroller, temperature sensor	NC	External	Phase-locked loop circuit (PLL)	0022 Hz	NC	Yes
[22]	Virtex 4, 32-bit MicroBlaze microprocessor	NC	External	Conventional frequency counter	<1 Hz	922 slices, 774 flip-flops, 1625 LUTs, 8 IOBs, 21 FIFO16, 1 GCLKS and 3 DSP48s	Yes
[30]	DE-2 board	PLL 200 MHz signal	External	Reciprocal counter, time-to-digital converters (TDCs)	0.25 Hz with reciprocal counter	NC	Yes
[31]	Arduino Atmega 2560 microcontroller, PIC16F628A per channel	Arduino 1 Hz signal	External	Conventional frequency counter	1 Hz	One microcon-troller per channel	No
[32]	Spartan 3	External 50 MHz oscillator	External	Conventional frequency counter	1 Hz	NC	Yes
[33]	CPLD XC2C256	16.9344 MHz external TCXO oscillator	External	Conventional frequency counter	0.5 Hz	228 macrocells, 19 function blocks and 72% of registers	Yes
[34]	Spartan-3E (XC3S250E)	100 MHz TCXO oscillator	External	Reciprocal counter	0.1 Hz	NC	Yes
Our QCR with reciprocal counter	Virtex-6 ML605	GPS 1PPS signal from MediaTek GPS chipset MT3339	Internal	Reciprocal counter	<0.1 Hz	188 slices, 733 flip-flops, 527 LUTs, 13 IOBs	Yes

5. Conclusions and Future work

This paper presents a GPS-based accurate multi-channel QCM sensor measurement application using an FPGA directly interfaced with quartz crystal oscillators. Our system ensures multi-sensor measurement while integrating the reciprocal frequency measurement part within the FPGA. The reconfigurable embedded system performs the frequency measurement for high-accuracy QCM applications which has been implemented and behavior-tested using the FPGA Virtex-6 XC6VLX240T. The proposed system requires low FPGA logic resources while taking advantage of the computation concurrency. Furthermore, the suggested system eliminates the need for external oscillators or chips for conditioning or data processing. It enables the measurement and monitoring of QCM frequencies within the 1–50 MHz range with an accuracy of 2 ppm and a precision of under 0.1 Hz. Frequency values from multiple QCM sensors can be measured and recorded periodically, every second, using this system.

As part of our future work, we also plan to integrate compensation conditions (or calibration sensors) based on Allan deviation computation within the FPGA. This integration will accurately determine the influence of measurement conditions on oscillator behavior, frequency stability, and sensor mass resolution, thereby improving accuracy and processing time [33]. Additionally, we will measure temperatures and ambient humidity in sensor environments simultaneously to enhance high-temperature compensation and improve quartz crystal characteristics. To achieve higher accuracy in frequency measurement, we will consider incorporating a time-to-digital converter IP core for FPGA, enabling precise measurements [34]. Moreover, as sensors with high sensitivity require more time for accurate frequency shift measurements, we will explore the principle of rational approximations for measurement [35].

Author Contributions: Conceptualization, A.B. and C.T.; Methodology, A.B.; Software, A.B.; Validation, C.T.; Resources, A.B.; Data curation, A.B.; Writing—original draft, A.B. and C.T.; Writing—review and editing, A.B. and C.T.; Supervision, C.D. and J.G.; Project administration, C.T. All authors have read and agreed to the published version of the manuscript.

Funding: This research received no external funding.

Data Availability Statement: The data presented in this study are available on request from the corresponding author. The data is not publicly available as it belongs to the LCOMS laboratory.

Conflicts of Interest: The authors declare no conflict of interest.

References

1. Zhan, M.; Xie, X. Precision Measurement Physics: Physics That Precision Matters. *Natl. Sci. Rev.* **2020**, *7*, 1795. [CrossRef] [PubMed]
2. Fauzi, F.; Rianjanu, A.; Santoso, I.; Triyana, K. Gas and Humidity Sensing with Quartz Crystal Microbalance (QCM) Coated with Graphene-Based Materials—A Mini Review. *Sens. Actuators A Phys.* **2021**, *330*, 112837. [CrossRef]
3. Burda, I. Advanced Impedance Spectroscopy for QCM Sensor in Liquid Medium. *Sensors* **2022**, *22*, 2337. [CrossRef] [PubMed]
4. Chen, W.; Deng, F.; Xu, M.; Wang, J.; Wei, Z.; Wang, Y. GO/Cu_2O Nanocomposite Based QCM Gas Sensor for Trimethylamine Detection under Low Concentrations. *Sens. Actuators B Chem.* **2018**, *273*, 498–504. [CrossRef]
5. Tan, J.; Xu, J. Applications of Electronic Nose (e-Nose) and Electronic Tongue (e-Tongue) in Food Quality-Related Properties Determination: A Review. *Artif. Intell. Agric.* **2020**, *4*, 104–115. [CrossRef]
6. Matsumoto, K.; Tiu, B.D.B.; Kawamura, A.; Advincula, R.C.; Miyata, T. QCM Sensing of Bisphenol A Using Molecularly Imprinted Hydrogel/Conducting Polymer Matrix. *Polym. J.* **2016**, *48*, 525–532. [CrossRef]
7. Clegg, J.R.; Irani, A.S.; Ander, E.W.; Ludolph, C.M.; Venkataraman, A.K.; Zhong, J.X.; Peppas, N.A. Synthetic Networks with Tunable Responsiveness, Biodegradation, and Molecular Recognition for Precision Medicine Applications. *Sci. Adv.* **2019**, *5*, eaax7946. [CrossRef] [PubMed]
8. Yakimova, L.S.; Ziganshin, M.A.; Sidorov, V.A.; Kovalev, V.V.; Shokova, E.A.; Tafeenko, V.A.; Gorbatchuk, V.V. Molecular Recognition of Organic Vapors by Adamantylcalix[4]Arene in QCM Sensor Using Partial Binding Reversibility. *J. Phys. Chem. B* **2008**, *112*, 15569–15575. [CrossRef]
9. Karczmarczyk, A.; Haupt, K.; Feller, K.-H. Development of a QCM-D Biosensor for Ochratoxin A Detection in Red Wine. *Talanta* **2017**, *166*, 193–197. [CrossRef]
10. Cervera-Chiner, L.; Juan-Borrás, M.; March, C.; Arnau, A.; Escriche, I.; Montoya, Á.; Jiménez, Y. High Fundamental Frequency Quartz Crystal Microbalance (HFF-QCM) Immunosensor for Pesticide Detection in Honey. *Food Control* **2018**, *92*, 1–6. [CrossRef]
11. Kuchmenko, T.A.; Lvova, L.B. A Perspective on Recent Advances in Piezoelectric Chemical Sensors for Environmental Monitoring and Foodstuffs Analysis. *Chemosensors* **2019**, *7*, 39. [CrossRef]
12. Sauerbrey, G. Verwendung von Schwingquarzen zur Wägung Dünner Schichten und zur Mikrowägung. *Z. Phys.* **1959**, *155*, 206–222. [CrossRef]
13. Yuwono, A.S.; Lammers, P.S. Odor Pollution in the Environment and the Detection Instrumentation. *Agric. Eng. Int.* **2004**, *VI*, 1–33.
14. Chen, Q.; Huang, X.; Yao, Y.; Mao, K. Analysis of the Effect of Electrode Materials on the Sensitivity of Quartz Crystal Microbalance. *Nanomaterials* **2022**, *12*, 975. [CrossRef] [PubMed]
15. Alassi, A.; Benammar, M.; Brett, D. Quartz Crystal Microbalance Electronic Interfacing Systems: A Review. *Sensors* **2017**, *17*, 2799. [CrossRef]
16. Park, J.-Y.; Choi, J.-W. Review—Electronic Circuit Systems for Piezoelectric Resonance Sensors. *J. Electrochem. Soc.* **2020**, *167*, 037560. [CrossRef]

17. Lucklum, R.; Eichelbaum, F. Interface Circuits for QCM Sensors. In *Piezoelectric Sensors*; Steinem, C., Janshoff, A., Eds.; Springer Series on Chemical Sensors and Biosensors; Springer: Berlin/Heidelberg, Germany, 2007; Volume 5, pp. 3–47. ISBN 978-3-540-36567-9.
18. Edvardsson, M.; Rodahl, M.; Kasemo, B.; Höök, F. A Dual-Frequency QCM-D Setup Operating at Elevated Oscillation Amplitudes. *Anal. Chem.* **2005**, *77*, 4918–4926. [CrossRef]
19. Eichelbaum, F.; Borngräber, R.; Schröder, J.; Lucklum, R.; Hauptmann, P. Interface Circuits for Quartz-Crystal-Microbalance Sensors. *Rev. Sci. Instrum.* **1999**, *70*, 2537–2545. [CrossRef]
20. Park, J.-Y.; Pérez, R.L.; Ayala, C.E.; Vaughan, S.R.; Warner, I.M.; Choi, J.-W. A Miniaturized Quartz Crystal Microbalance (QCM) Measurement Instrument Based on a Phase-Locked Loop Circuit. *Electronics* **2022**, *11*, 358. [CrossRef]
21. Fang, Y. Design of Equal Precision Frequency Meter Based on FPGA. *Engineering* **2012**, *4*, 696–700. [CrossRef]
22. Moure, M.J.; Valdes, M.D.; Rodiz, P.; Rodriguez-Pardo, L.; Farina, J. An FPGA-Based System on Chip for the Measurement of QCM Sensors Resolution. In Proceedings of the 2006 International Conference on Field Programmable Logic and Applications, Madrid, Spain, 28–30 August 2006; IEEE: Madrid, Spain, 2006; pp. 1–4.
23. Valdes, M.D.; Moure, M.J.; Rodriguez, L.; Farina, J. Implementation of a Frequency Measurement Circuit for High-Accuracy QCM Sensors. In Proceedings of the 2008 4th Southern Conference on Programmable Logic, Bariloche, Argentina, 26–28 March 2008; IEEE: Bariloche, Argentina, 2008; pp. 233–236.
24. Xilinx, Datasheet Virtex-6 FPGA ML-605 Evaluation Kit, Part Number: EK-V6-ML605-G. 2012. Available online: https://docs.xilinx.com/v/u/en-US/ug534 (accessed on 2 February 2023).
25. Osterdock, T.N.; Kusters, J.A. Using a New GPS Frequency Reference in Frequency Calibration Operations. In Proceedings of the 1993 IEEE International Frequency Control Symposium, Salt Lake City, UT, USA, 2–4 June 1993; IEEE: Salt Lake City, UT, USA, 1993; pp. 33–39.
26. Stovbun, N.S.; Gulyakov, S.A. Development of the Dual-Channel Frequency Meter for Measurements with Hydrostatic Pressure Sensor. *IOP Conf. Ser. Earth Environ. Sci.* **2021**, *946*, 012018. [CrossRef]
27. OBUFDS. Vivado Design Suite 7 Series FPGA and Zynq-7000 SoC Libraries Guide (UG953) • Reader • Documentation Portal. Available online: https://docs.xilinx.com/r/en-US/ug953-vivado-7series-libraries/OBUFDS (accessed on 2 February 2023).
28. Valdes, M.D.; Moure, M.J.; Rodriguez, L.; Farina, J. Improving a Frequency Measurement Circuit for High-Accuracy QCM Sensors. In Proceedings of the 2008 IEEE International Symposium on Industrial Electronics, Cambridge, UK, 30 June–2 July 2008; IEEE: Cambridge, UK, 2008; pp. 998–1002.
29. Batarseh, M.G.; Al-Hoor, W.; Huang, L.; Iannello, C.; Batarseh, I. Window-Masked Segmented Digital Clock Manager-FPGA-Based Digital Pulsewidth Modulator Technique. *IEEE Trans. Power Electron.* **2009**, *24*, 2649–2660. [CrossRef]
30. Fernandez Molanes, R.; Farina, J.; Rodriguez-Andina, J.J. Field-Programmable System-on-Chip for High-Accuracy Frequency Measurements in QCM Sensors. In Proceedings of the IECON 2013—39th Annual Conference of the IEEE Industrial Electronics Society, Vienna, Austria, 10–13 November 2013; IEEE: Vienna, Austria, 2013; pp. 2267–2272.
31. Karapınar, M.; Gürkan, S.; Öner, P.A.; Doğan, S. Design of a Multi-Channel Quartz Crystal Microbalance Data Acquisition System. *Meas. Sci. Technol.* **2018**, *29*, 075009. [CrossRef]
32. Misbah, M.; Rivai, M.; Kurniawan, F. Quartz Crystal Microbalance Based Electronic Nose System Implemented on Field Programmable Gate Array. *TELKOMNIKA* **2019**, *17*, 370. [CrossRef]
33. Wijayanto, V.R.; Sakti, S.P. Design of Dual Edge 0.5 Hz Precision Frequency Counter for QCM Sensor. *Appl. Mech. Mater.* **2015**, *771*, 29–32. [CrossRef]
34. Syahbana, M.A.; Santjojo, D.J.H.D.; Sakti, S.P. High-Resolution Multiple Channel Frequency Counter Using Spartan-3E FPGA. In Proceedings of the 2016 International Seminar on Sensors, Instrumentation, Measurement and Metrology (ISSIMM), Malang, Indonesia, 10–11 August 2016; IEEE: Malang, Indonesia, 2016; pp. 111–114.
35. Murrieta-Rico, F.N.; Petranovskii, V.; Sergiyenko, O.Y.; Hernandez-Balbuena, D.; Lindner, L. A New Approach to Measurement of Frequency Shifts Using the Principle of Rational Approximations. *Metrol. Meas. Syst.* **2017**, *24*, 45–56. [CrossRef]

Disclaimer/Publisher's Note: The statements, opinions and data contained in all publications are solely those of the individual author(s) and contributor(s) and not of MDPI and/or the editor(s). MDPI and/or the editor(s) disclaim responsibility for any injury to people or property resulting from any ideas, methods, instructions or products referred to in the content.

electronics MDPI

Article

Improving Seed-Based FPGA Packing with Indirect Connection for Realization of Neural Networks

Le Yu [1,*], Baojin Guo [1], Tian Zhi [2] and Lida Bai [3]

[1] School of Artificial Intelligence, Beijing Technology And Business University, Beijing 100048, China; 2030602043@st.btbu.edu.cn
[2] Institute of Computing Technology, Chinese Academy of Sciences, Beijing 100190, China; zhitian@ict.ac.cn
[3] Shandong Cwise Microelectronics Technology Co., Ltd., Jinan 250102, China; peada@126.com
* Correspondence: yule@btbu.edu.cn

Abstract: FPGAs are gaining favor among researchers in fields including artificial intelligence and big data due to their configurability and high level of parallelism. As the packing methods indisputably affect the implementation performance of FPGA chips, packing techniques play an important role in the design automation flow of FPGAs. In this paper, we propose a quantitative rule for packing priority of neural network circuits, and optimize the traditional seed-based packing methods with special primitives. The experiment result indicates that the proposed packing method achieves an average decrease of 8.45% in critical path delay compared to the VTR8.0 on Koios deep learning benchmarks.

Keywords: FPGA; EDA; packing; seed-based; indirect connections

Citation: Yu, L.; Guo, B.; Zhi, T.; Bai, L. Improving Seed-Based FPGA Packing with Indirect Connection for Realization of Neural Networks. *Electronics* **2023**, *12*, 2691. https://doi.org/10.3390/electronics12122691

Academic Editors: Akash Kumar, Andres Upegui, Laurent Gantel and Andrea Guerrieri

Received: 11 April 2023
Revised: 10 June 2023
Accepted: 14 June 2023
Published: 15 June 2023

Copyright: © 2023 by the authors. Licensee MDPI, Basel, Switzerland. This article is an open access article distributed under the terms and conditions of the Creative Commons Attribution (CC BY) license (https://creativecommons.org/licenses/by/4.0/).

1. Introduction

In recent years, Field Programmable Gate Array (FPGA) chips are being widely used in the acceleration of neural networks (NNs). NN applications such as image classification [1], object detection [2], and natural language processing [3] can take full advantage of the reconfigurable parallelism of FPGA architectures.

FPGAs typically consist of two-dimensional reconfigurable arrays, including Configurable Logic Blocks (CLBs), Block RAMs (BRAMs), Digital Signal Processing blocks (DSPs) [4,5], etc., and all these tiles are connected through programmable wires and switches. The back-end optimization of FPGAs is restricted by the pre-placed computing primitives and the pre-routed clock tree. As the packing methods indisputably affect the implementation performance of FPGA chips, packing techniques play an important role in the design automation flow of FPGAs.

Nowadays, the most commonly used packing algorithms are Seed-based algorithms, which pack the look up tables (LUTs) and flip-flops (FFs) together to implement the designated logic function. Seed-based algorithms construct new tiles by seeding each with an unpacked primitive and greedily absorbing its surrounding primitives according to attraction functions. However, emerging heterogeneous FPGA architectures present a new challenge to the traditional packing methods; heterogeneous IP Blocks, such as the BRAMs and DSPs, make traditional packing methods inefficient due to the unevenly distributed wiring topology.

In this paper, we propose an improved packing algorithm. The main contributions of this paper are as follows: (1) A quantitative rule for packing priority of neural network circuits is proposed. (2) The traditional seed-based packing methods with special primitives is optimized. Compared with Verilog-To-Routing(VTR) [6], the proposed packing method achieves an average reduction of 8.45% in latency at the cost of a 0.58% increase in resource consumption and a 7.55% increase in runtime for the optimized circuits without affecting other circuits.

2. Related Work

Previous algorithms for FPGA packing can be loosely categorized into seed-based packing and partitioning-based packing.

VPACK [7] is the first seed-based packing approach. It packs LUTs and FFs into BLE, and then into CLBs. T-VPACK [8] reduces the critical path delay of the circuit by modifying the attraction function. DPACK [9] adds the Manhattan distance to the attractive function of T-VPACK, which reduces the bus length by 16% and the critical path delay by 8% after placement and routing. For more complex logic blocks, ref. [10] proposes the AAPACK algorithm, which packs primitives into molecules and assemble clusters from a set of molecules. RSVPACK [11] is a packing algorithm for the XILINX V6 architecture that bridges the gap between academia and industry. DPPACK [12] adopts distributed parallel packing, which shortens the runtime by 1.4–3.2 times with acceptable quality degradation compared to AAPACK. [6] is an update to AAPack, optimized for seed selection and attraction functions.

PPFF [13] applies partition-based packing to FPGAs as a sub-step of placement. PPack [14] explores partition-based packing, which adds a significant amount of runtime compared to T-VPACK. PPack2 [15] is an improved version of PPack. Compared to T-VPack, PPack2 has an 11.2% reduction in critical path delay. PartSA [16] is a multi-threaded parallel packing algorithm that reduces runtime but increases wire length by 26%.

A summary of FPGA packing algorithms is shown in Table 1. Partitioning-based algorithms are effective at packing simple FPGAs, but they can struggle to handle the constraints present in commercial devices. Conversely, seed-based algorithms perform better in packing heterogeneous FPGAs. The seed-based algorithm adopts the same packing rule for tiles with different areas, which will affect the wire length around some tiles and even the delay of the critical path. This is more prominent in neural network circuits. This paper proposes improved seed-based packing algorithms for neural networks, which can reduce the critical path delays in the packing process.

Table 1. Summary of FPGA Packing Methods.

Packing Methods	Advantages	Disadvantages
partitioning-based packing [13–16]	effective	struggle to handle packing of heterogeneous clusters
seed-based packing [6–12]	excel at packing heterogeneous FPGAs	the influence of tiles of different sizes on wirelength was ignored

3. User Netlist

After circuit synthesis and mapping, a primitive-level user netlist is generated. It is composed of primitives and necessary connectivity.

3.1. Primitives

Primitives are fundamental units that cannot be separated, and their internal structure is usually treated as a black box. Each primitive contains one or more ports, including input and output ports, with each port having one or more pins. For example, the input port of a LUT usually has four to eight pins. In modern FPGAs, three frequently used primitives are DSPs, RAMs, and adders, which were not present in earlier generations of FPGA products. The features of these three types of primitives in application circuits are described below.

3.1.1. DSPs and RAMs

DSPs and RAMs are embedded reconfigurable IP blocks in FPGAs. However, the use of primitives inherently introduces significant latency. In advanced neural networks, about 80–90% of the operations are matrix multiplication [17], which means that the critical path often includes DSPs or RAMs. Consequently, reducing the delay of the network connected by these primitives has become a pressing issue that requires immediate attention in the field.

3.1.2. Adders

Adders typically implement carry chains and are generally located in CLBs. In real-world applications, adders typically consist of multiple cascaded CLBs. Since adders re-use the LUT routing pins, a smaller number of LUT pins is required for CLBs with adders. Taking the Intel Stratix_10 architecture as an example, a CLB without adders can absorb a 6-bit LUT, while for CLBs with adders, the maximum number of input pins is four.

3.2. Connectivity

The connectivity between primitives in the user netlist is implemented through networks. Primitives are connected to a network through pins. A connected network is usually interconnected to the pins of multiple primitives, among which only one pin is an output pin. A network with a large number of primitives connected to it is commonly referred to as a high fan-out network, while a network with a smaller number of primitives is typically known as a low fan-out network.

There are three types of connectivity between primitives: direct connectivity, indirect connectivity and high fan-out connectivity. Direct connectivity refers to the connectivity between primitives through a low fan-out network. Indirect connectivity is the connection between primitives through different networks of the same tile. A high fan-out connectivity refers to the connectivity between primitives through a high fan-out network. The modes of connection are shown in Figure 1.

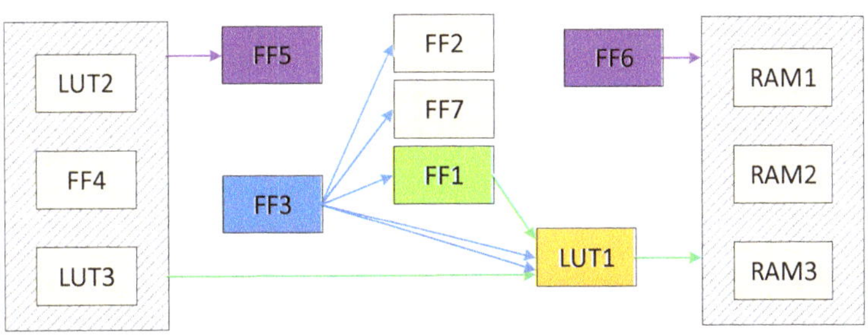

Figure 1. There are three types of connectivity between primitives. LUT1 is the seed, FF1 is a directly connected primitive, FF5 and FF6 are indirectly connected primitives, and FF3 is a high fan-out connected primitive.

Packer is designed to absorb the primitives of direct connections in order to reduce the number of external networks on the tile, which serves to reduce the overload of routing and computing. For primitives of indirect connections, packing them into the tiles can reduce the number of external connections of the tiles, which increases the relevancy of the connected tile.

4. Packing Methods

The path delay of FPGAs is affected by two factors: the internal delay of logic blocks and the delay introduced by programmable routing. The delay of logic blocks is fixed, while the delay of programmable routing is influenced by the wire length of nets. In order to minimize the delay of the nets that are connected to DSPs and RAMs, it is essential to reduce the length of these nets. However, since the area occupied by DSPs and RAMs is much larger than that of CLBs, some pins on DSPs and RAMs may be located far apart. As a result, even tiles that are situated around a DSP or RAM may be distant from the pins of the corresponding connection. Figure 2 shows the result of placement and routing in VTR8.0. In the figure, o1 stands for the DSP, o2 stands for the CLB. And two primitives indirectly connected through o1 are grouped in o2, resulting in two nets between o1 and o2. Figure 2a shows two routing networks connected with DSP pins at the same coordinates, while Figure 2b shows two routing networks connected with DSP pins at different coordinates. It can be seen from Figure 2 that absorbing primitives indirectly connected through pins in the same location can reduce wire length. This is in contrast to absorbing primitives indirectly connected through pins in different locations. To minimize the delay of the nets that are linked to DSPs and RAMs, the packer prioritizes absorbing the primitives that are indirectly connected through pins in the same location based on the pin distribution of tile. This approach avoids absorbing primitives that are indirectly connected at different locations, thereby reducing the length of the nets connected by DSP and RAM after placement and routing.

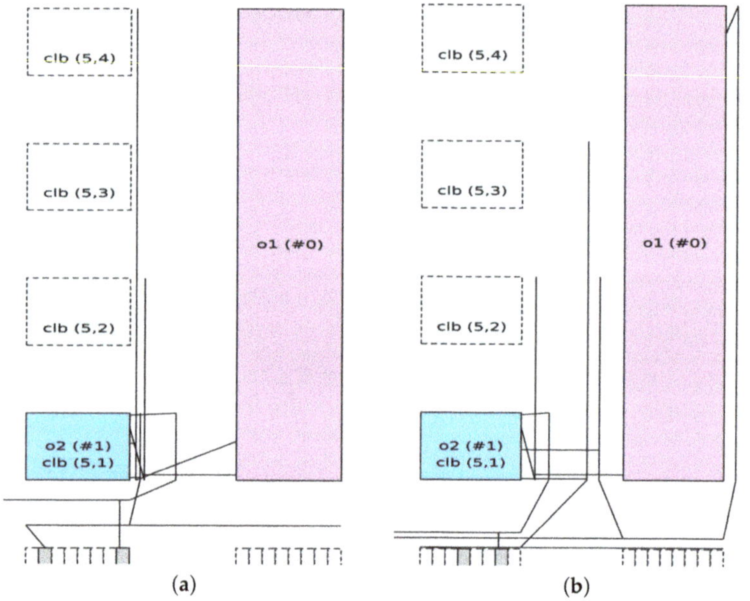

Figure 2. Connections between a CLB and a DSP via two nets (VTR8.0 visualization).

When a CLB is connected to multiple DSPs or RAMs, the average wire length between the CLB and these tiles tends to be high, as shown in Figure 3. This issue becomes more prevalent if there is a higher proportion of RAMs and DSPs in the circuit, as it increases the likelihood of multiple connections between the same CLB and these tiles. In contrast, if the circuit design incorporates a high proportion of adders, there will be fewer options for packers, and primitives in different positions will be absorbed. Moreover, the cascading of adders considers multiple CLBs as a single unit, which is often connected to multiple DSPs or RAMs. This interconnection can significantly impact the overall packing result.

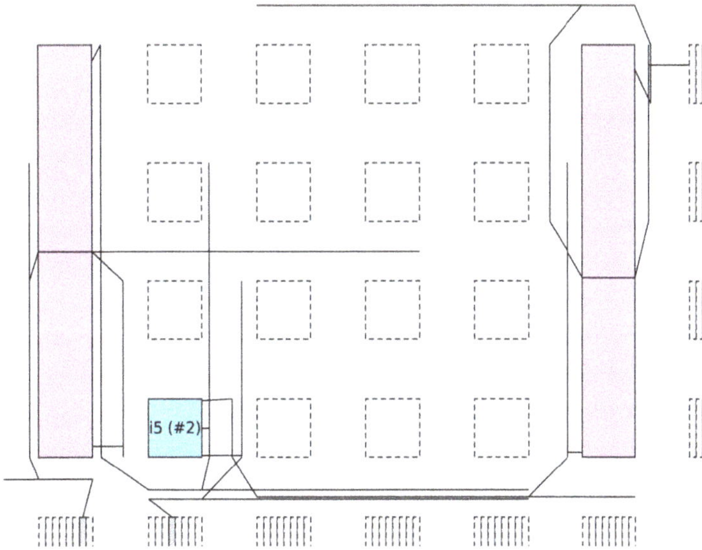

Figure 3. A CLB connects multiple DSPs (VTR8.0 visualization).

Table 2 presents a comparison between wire lengths and wiring segments utilized in connecting DSPs and CLBs. As the table illustrates, the connection relationships between CLBs, DSPs, and RAMs significantly impact wire length and wiring segment consumption. Therefore, it is essential to give priority to circuits that are indirectly connected via DSPs and RAMs. In this study, we refer to DSPs and RAMs that satisfy the specified requirements as special primitives, while referring to other primitives as normal primitives.

Table 2. Three types of connections between DSP and CLB.

Types of Connections	Total Wirelength	Maximum Net Length	Total Wiring Segments Used	Maximum Segments Used by a Net
Figure 2a	14	5	6	2
Figure 2b	21	9	8	3
Figure 3	47	12	17	5

The process of the packing algorithm in this paper is shown in Algorithm 1. First, the packer analyzes the proportion of various primitives in the user netlist to determine whether to use DSP and RAM as special primitives or not. The packer then groups the primitives into molecules, calculates the seed gain for each molecule and selects the molecule with the highest gain as the seed. After the seed is selected, the packer will absorb the molecules around the tile until the constraints of the tile are no longer satisfied or the surrounding molecules are all packed. The above process is repeated until all molecules are packed. Our packing algorithm is composed of three stages: primitive classification, seed selection, and molecule selection.

Algorithm 1 Pack algorithm.

Input: $fpga\_architecture$ and $netlist$
Output: $packed\_netlist$

```
if number(DSP, RAM, adder) < threshold then
    DSP, RAM ∈ special tiles
else
    DSP, RAM ∈ normal tiles
end if
moles ← atom2molecule(netlist)
while unpacked_mol ≠ ∅ do
    seed ← get_highest_seed_gain(moles)
    cur_clus ← creat_cluster(seed)
    while have_spcace(cur_clus) do
        next_mol ← get_highest_gain(moles)
        if next_mol = ∅ then
            break
        end if
        cur_clus ← add_mole(moles, cur_clus)
    end while
end while
return packed_netlist
```

4.1. Primitive Classification

In this paper, the algorithm considers the proportion of DSPs, RAMs, and adders in the circuit as the quantitative rule for special primitives. The formula used for this purpose is as follows:

$$SP = \begin{cases} \{DSPs, RAMs\}, & \dfrac{num(DSPs) + num(RAMs) + num(adders)}{num(total)} < thre \\ \varnothing, & \text{otherwise} \end{cases} \tag{1}$$

where SP is a set of special primitives, $num(DSPs)$ is the number of DSPs in the netlist, $num(RAMs)$ is the number of RAMs in the netlist, $num(adders)$ is the number of adders in the netlist, $num(total)$ is the total number of primitives in the netlist, and $thre$ is the threshold.

4.2. Seed Selection

The selection of the seed impacts the order in which different parts of the netlist will be clustered. In VTR8.0, the criteria of the selection of seed are determined by the number of primitives in the molecule, the number of molecular pins, and the delay information of the molecule. When the molecules are packed, the packer can determine the type of connectivity between the pins of the tile and the network. In this paper we seek to raise the priority of molecules with special primitives as seeds through the use of *seed_gain* as the criterion for seed selection. The molecule with large *seed_gain* is preferentially selected as the seed. The model of *seed_gain* is as follows:

$$\begin{aligned} seed\_gain &= w1 \times num(in) + w2 \times num(used\_in) \\ &+ w3 \times num(block) + w4 \times crit + w5 \times i(SP) \end{aligned} \tag{2}$$

where $num(used\_in)$ is the normalized number of input pins used, $num(in)$ is the normalized number of input pins, and $num(block)$ is the normalized number of primitives in the molecule, $crit$ is the delay of the primitive pins, $i(SP)$ is used to determine whether the current primitive is a special primitive, $w1 \sim w5$ is the weight.

4.3. Molecule Selection

Once a seed molecule has been chosen and a new tile is opened, the packer begins searching for unclustered molecules to add. The next molecule to be grouped into the current tile is determined by attraction functions, which are influenced by the connectivity between the molecule and the current tile.

For the attraction function of direct connectivity, our packing algorithm adopts the same attraction function as VTR8.0.

$$Aff(p,B) = (1-\beta) \times c\_gain(p,B) + \beta \times t\_gain(p,B) \tag{3}$$

where $c\_gain(p,B)$ is the connection benefit of the molecule p to the tile B, and $t\_gain(p,B)$ is the Criticality of the network connection between p and B. $c\_gain(p,B)$ formula is as follows.

$$c\_gain(p,B) = \frac{(1-\alpha) \times nets(p,B) + \alpha \times con(p,B)}{num\_pins(p)} \tag{4}$$

where $nets(p,B)$ is the number of shared nodes between the molecule p and the tile B, and $con(p,B)$ and the pins of p are closely related to the connection relationship of B; the formula is as follows.

$$con(p,B) = \frac{1}{ext(p,B) + packed(p) + 1} \tag{5}$$

where $ext(p,B)$ is the number of pins of p that are not connected to tile B, and $packed(p)$ is the number of pins of p that are connected to the packed molecule.

Our packing algorithm considers two types of indirect connectivity: indirect connectivity through special tiles and indirect connectivity through normal tiles. The packer should preferentially absorb molecules that are indirectly connected to the current tile via short-distance pins. Therefore, the molecules that are indirectly connected to the current tile through the special tile are divided into three categories. The first type is molecules that are indirectly connected through pins on the same side and at the same location, as shown in Figure 4a. The second type is molecules that are indirectly connected on the same side but at different locations, as shown in Figure 4b. The third type is molecules indirectly connected by pins on different sides, as shown in Figure 4c. During packing, the packer prioritizes first-type molecules into the same tile, and then considers second-type molecules, and finally third-type molecules. The cost of the attractive function is as follows.

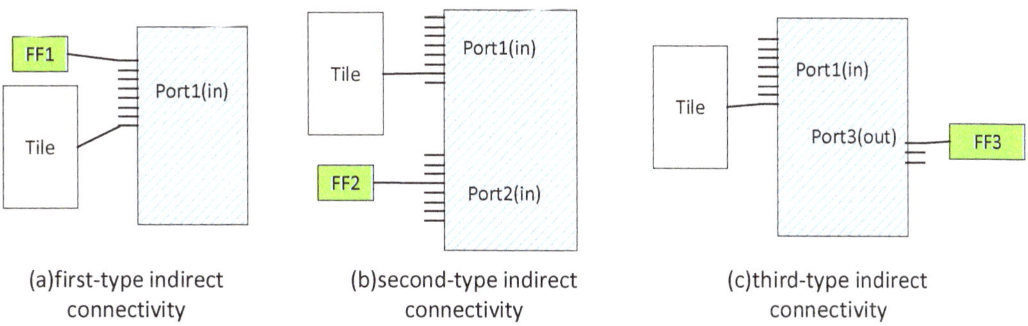

Figure 4. Three models of indirect connectivity.

$$Aff(p,B) = \sum_{T_i \in Tiles} ind\_gain(p,B,T_i) \tag{6}$$

Among them, *Tiles* is the set of tiles around tile B. $ind\_gain(p, B, T_i)$ is the attraction of p indirectly connected to molecule tile B through Ti.

$$ind\_gain(p, B, T_i) = \begin{cases} w_{port} \times num_{port} + w_{dir} \times num_{dir} + w_{rev} \times num_{rev}, T_i \in SP \\ w_{nor} \times n_{nor}, \quad \text{otherwise} \end{cases} \tag{7}$$

Among them, w_{port} is the weight of the first type of molecules, w_{dir} is the weight of the second type of molecules, w_{rev} is the weight of the third type of molecules, num_{port}, num_{dir} and num_{rev} are the connection times of the three indirect connection molecules, w_{nor} is the weight of molecules indirectly connected through normal tiles, and n_{nor} is the number of times primitives are indirectly connected through normal tiles. The formula for w_{dir} is as follows:

$$w_{dir} = \frac{w_{port}}{n_{dir}} \tag{8}$$

Among them, n_{dir} is a positive integer.

If both directly connected and indirectly connected molecules are packed, and the tile to be packed does not meet the constraints, then the molecules connected by the high fan-out network are selected for clustering.

5. Experimental Results

The experiments are performed on a workstation with an AMD EPYC 7302P (16 cores, 3 GHz) with 64 G of memory. The FPGA architecture used in this paper is the k6FracN10LB _mem20K_complexDSP_customSB_22nm architecture provided by VTR. Its blocks are Agilex-like, but the routing architecture is Stratix-IV-like [18]. The circuits used in this paper are from the Koios benchmark [19]. The Koios benchmark contains 20 deep learning-related circuits, all of which are medium- or large-size circuits, suitable for architecture research and EDA algorithm research. This paper runs Koios with a channel width of 200 for medium-size circuits and 300 for large-size circuits.

Table 3 shows some parameters and parameter values used in this experiment, and the values are obtained through verification in VTR8.0.

Table 3. Values of parameters.

Parameters	Value
$w1$	0.5
$w2$	0.2
$w3$	0.2
$w4$	0.1
w_{nor}	0.003
w_{rev}	0.001
α	0.6
β	0.2

Figure 5 shows the impact of the proportion of DSPs, RAMs and adder on the critical path delay in the Koios benchmark. It can be seen from Figure 5 that when the proportion of DSPs, RAMs and adder in the circuit exceeds 20%, the packer uses DSPs and RAMs as special primitives to cluster. This clustering results in a higher possibility of increasing the critical path delay, as shown in the red zone on the left in Figure 5. If the proportion of DSP in the circuit is less than 20%, eleven out of twelve benchmark circuits have successfully reduced critical path delays, as shown in the blue zone on the right in Figure 5. So the algorithm in this paper sets the *thre* as 20%.

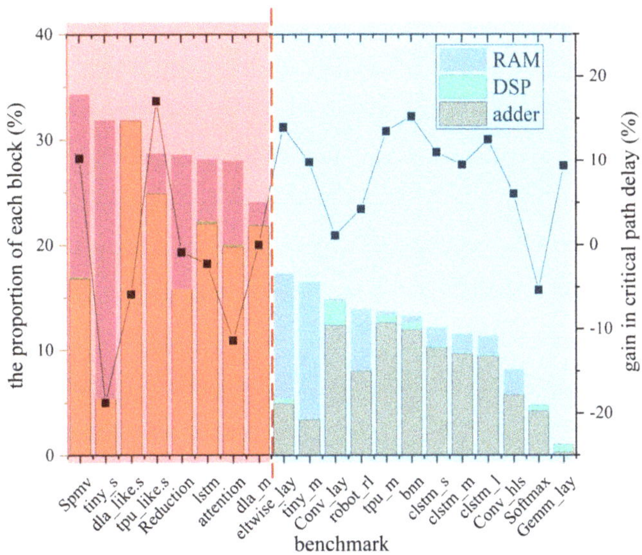

Figure 5. The influence of the proportion of DSPs, RAMs and adder in the circuit on the critical path delay. The histogram is the proportion of DSPs, RAMs and adder in the circuit, corresponding to the coordinates on the left. The line graph is the optimization rate of critical path delay, corresponding to the coordinates on the right.

In the seed selection stage, the effect of different w5 in the *seed_gain* on the algorithm of this paper was tested. For testing purposes, the medium circuits of the Koios benchmark that meet the special primitive conditions are used as the test circuits, and the results of these tests are shown in Figure 6. From the figure, it can be seen that the critical path delay is optimized best when w5 is 0.1. In this paper, we set w5 to 0.1.

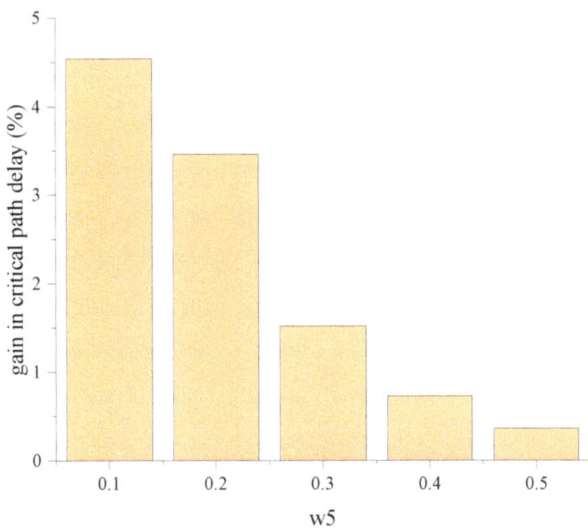

Figure 6. Effect of variation of w5 in seed selection stage on critical path delay.

In the molecule selection stage, the attraction of directly connected molecules should be greater than that of indirectly connected molecules. The attraction functions of indirectly connected molecules are shown in Equations (7) and (8). In indirect connection, the attraction should meet the condition $w_{port}>max\{w_{dir}, w_{nor}\}$ and $min\{w_{dir}, w_{nor}\}> w_{rev}$. With the parameters of Table 3, the attraction of direct connection molecules is greater than 0.1. Therefore, this paper sets w_{port} as 0.009, 0.03, 0.06 and 0.09, respectively, and observes the impact of changes in w_{port} on the critical path delay. As can be seen from Figure 7, in the Koios benchmark, the circuits that meet the special primitive conditions achieve better results when w_{port} is 0.03 for critical path delay.

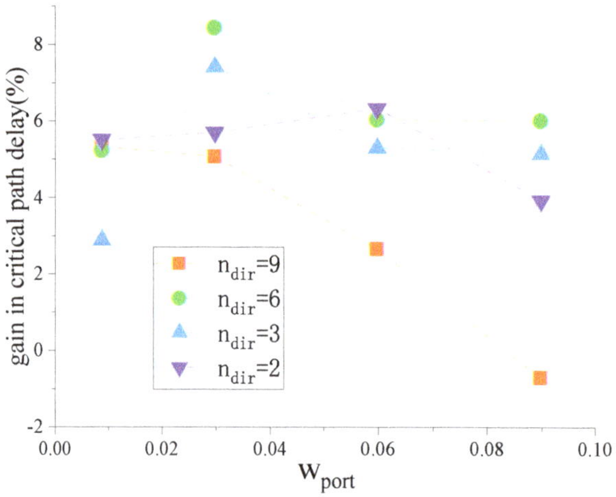

Figure 7. The impact of changes in w_{port} on critical path delays.

When dealing with molecules indirectly connected through special tiles, the first type of indirect connectivity should exhibit greater attraction than the second type. However, if the attraction of the second type of indirectly connected molecules is too small, the packer may end up neglecting these molecules and instead absorb those that are indirectly connected through another special tile. In this paper, n_{dir} is set to 2, 3, 6, and 9 respectively, and the critical path delay changes are observed. It can be seen from Figure 8 that in the Koios benchmark, the circuits that meet the special primitive conditions achieve better results in critical path delay when n_{dir} is 6.

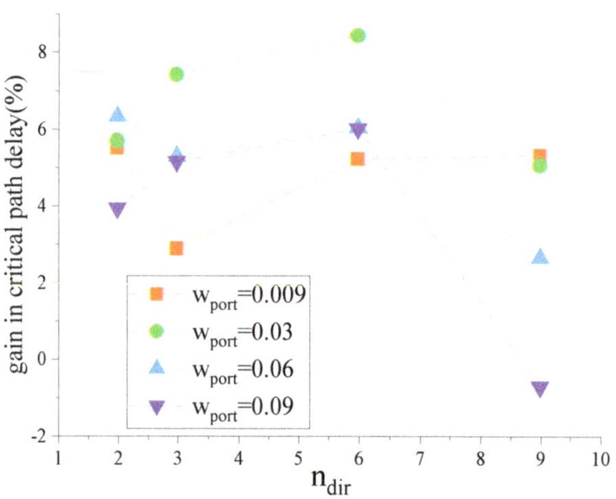

Figure 8. The impact of changes in n_{dir} on critical path delay.

From Figures 9 and 10, it can be inferred that resource consumption is scarcely affected by variations in w_{port} and n_{dir}. The variation range is within 0.24%. After rebalancing resource consumption and critical path delay, this paper sets w_{port} to 0.03 and n_{dir} to 6.

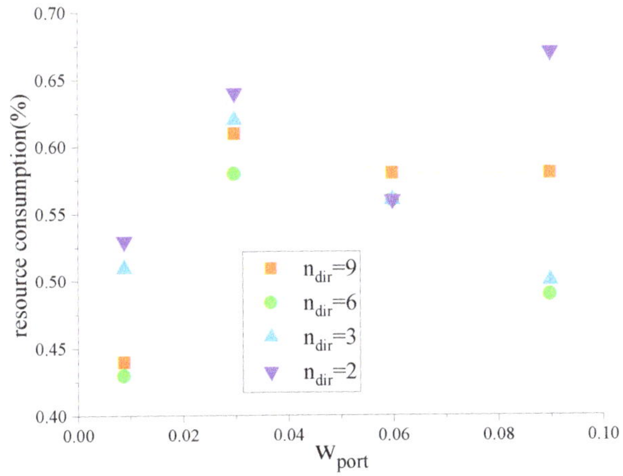

Figure 9. The impact of changes in w_{port} on resource consumption.

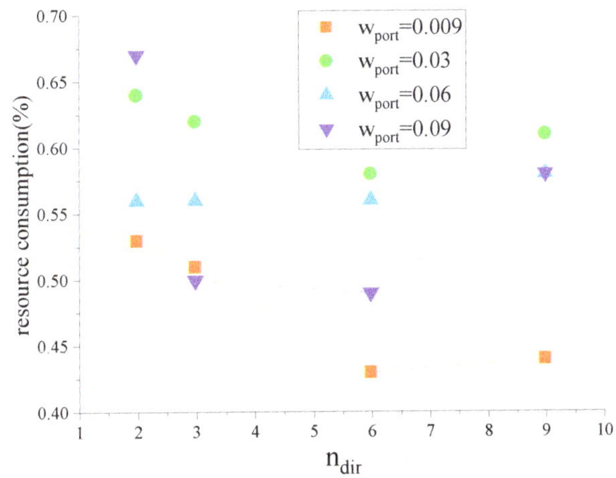

Figure 10. The impact of changes in n_{dir} on resource consumption.

Our proposed algorithm contrasts with the packing algorithm in VTR8.0 through a modified packing rule for indirectly connected molecules. This modification results in a rise in computational demand and extends the algorithm's runtime. However, while fewer options during the packing process translate into a slight increase in resource consumption, the algorithm's refinement leads to a shortened wirelength for nets around DSPs and RAMs. The end result is a reduction in the critical path delay.

As can be seen from Table 4, compared with VTR8.0, our packing reduces the critical path delay by 8.45% on average at the cost of a 0.58% increase in resource consumption and a 7.55% increase in runtime. Among the circuits in the Koios benchmark suite that meet the special primitive criteria, eleven out of twelve have successfully reduced critical path delays. For circuits that do not meet the conditions of special primitives, the algorithm in this paper does not divide the primitives around DSP and RAM, so resource consumption and critical path delay are the same as VTR8.0.

Table 4. The comparison between the packing method of the present invention and the results of VTR8.0 after placement and routing.

Circuits	Blocks		Runtime		Crit Path Delay	
	VTR8.0	Ours	VTR8.0	Ours	VTR8.0	Ours
Eltwise_layer.v	478	478	6.48	6.08	6.69	5.75
Conv_layer.v	1323	1326	42.47	43.24	6.9	6.82
Softmax.v	570	570	10.13	10.44	8.62	9.08
Gemm_layer.v	2217	2217	17.54	17.73	6.05	5.48
Robot_rl.v	1438	1443	19.19	19.46	12.66	12.12
Conv_layer_hls.v	1749	1748	115.18	137.12	6.9	6.48
bnn.v	1409	1452	378.2	575.76	7.98	6.76
Tiny_darknet.med.v	18,315	18,380	3164.24	3890.84	13.68	12.33
Tpu_like.medium.v	5344	5436	787.39	682.35	12.34	10.67
Clstm_like.small.v	10,078	10,130	308.05	326.07	8.06	7.17
Clstm_like.med.v	19,087	19,172	617.4	629.25	9.22	8.34
Clstm_like.large.v	28,118	28,232	985.77	992.84	10.51	9.19
Average	1.000	1.0058	1.00	1.0755	1.00	0.9155
improve	−0.58%		−7.55%		8.45%	

6. Conclusions and Future Work

This paper proposes a packing algorithm for FPGA improved by indirect connection, which refines the packing guideline in two aspects. (1) It proposes the quantitive rules of the special primitives by the proportion of DSPs, RAMs and adders. (2) It optimizes the traditional seed-based packing methods with special primitives, such as the modified criteria for seed and molecule selection. For circuits with special primitives, the proposed packing algorithm reduces the critical path delay by an average of 8.45% compared to VTR8.0. For circuits without special primitives, the critical path delay of our packing is the same as that of VTR8.0.

For future work, we will primarily aim at utilizing parallel computing methods to curtail the runtime of our proposed algorithm. We also plan to investigate the feasibility of porting these algorithms to commercial FPGAs.

Author Contributions: Conceptualization, L.Y. and B.G.; methodology, L.Y. and B.G.; software, B.G. and L.B.; validation, L.Y. and B.G.; formal analysis, B.G.; writing—original draft preparation, B.G.; writing—review and editing, L.Y. and T.Z.; supervision, L.Y. All authors have read and agreed to the published version of the manuscript.

Funding: This research received no external funding.

Data Availability Statement: The benchmark and architecture files used in this paper are both from VTR (Verilog-to-Routing) https://github.com/verilog-to-routing/vtr-verilog-to-routing.

Conflicts of Interest: The authors declare no conflict of interest.

References

1. Ghani, A.; Hodeify, R.; See, C.H.; Keates, S.; Lee, D.J.; Bouridane, A. Computer Vision-Based Kidney's (HK-2) Damaged Cells Classification with Reconfigurable Hardware Accelerator (FPGA). *Electronics* **2022**, *11*, 4234. [CrossRef]
2. Zhang, N.; Wei, X.; Chen, H.; Liu, W. FPGA Implementation for CNN-Based Optical Remote Sensing Object Detection. *Electronics* **2021**, *10*, 282. [CrossRef]
3. Han, Z.; Jiang, J.; Qiao, L.; Dou, Y.; Xu, J.; Kan, Z. Accelerating Event Detection with DGCNN and FPGAs. *Electronics* **2020**, *9*, 1666. [CrossRef]
4. Betz, V.; Rose, J.; Marquardt, A. *Architecture and CAD for Deep-Submicron FPGAs*; Springer: Boston, MA, USA, 1999; pp. 11–18.
5. 7 Series FPGAs Configurable Logic Block User Guide. Available online: https://docs.xilinx.com/v/u/en-US/ug474_7Series_CLB (accessed on 9 June 2023).
6. Murray, K.E.; Petelin, O.; Zhong, S.; Wang, J.M.; Eldafrawy, M.; Legault, J.P.; Sha, E.; Graham, A.G.; Wu, J.; Walker, M.J.P.; et al. VTR 8: High-Performance CAD and Customizable FPGA Architecture Modelling. *ACM Trans. Reconfig. Technol. Syst.* **2020**, *13*, 1936–7406. [CrossRef]

7. Betz, V.; Rose, J. Cluster-based logic blocks for FPGAs: Area-efficiency vs. input sharing and size. In Proceedings of the Custom Integrated Circuits Conference (CICC 97), Santa Clara, CA, USA, 5–8 May 1997; pp. 551–554. [CrossRef]
8. Marquardt, A.S.; Betz, V.; Rose, J. Using Cluster-Based Logic Blocks and Timing-Driven Packing to Improve FPGA Speed and Density. In Proceedings of the 1999 ACM/SIGDA Seventh International Symposium on Field Programmable Gate Arrays, Monterey, CA, USA, 21–23 February 1999; pp. 37–46. [CrossRef]
9. Chen, D.T.; Vorwerk, K.; Kennings, A. Improving Timing-Driven FPGA Packing with Physical Information. In Proceedings of the 2007 International Conference on Field Programmable Logic and Applications, Amsterdam, The Netherlands, 27–29 August 2007; pp. 117–123. [CrossRef]
10. Luu, J.; Anderson, J.H.; Rose, J.S. Architecture Description and Packing for Logic Blocks with Hierarchy, Modes and Complex Interconnect. In Proceedings of the 19th ACM/SIGDA International Symposium on Field Programmable Gate Arrays, Monterey, CA, USA, 27 February–1 March 2011; pp. 227–236. [CrossRef]
11. Haroldsen, T.; Nelson, B.; Hutchings, B. Packing a modern Xilinx FPGA using RapidSmith. In Proceedings of the 2016 International Conference on ReConFigurable Computing and FPGAs (ReConFig), Cancun, Mexico, 30 November–2 December 2016; pp. 1–6. [CrossRef]
12. Chen, Q.; Shen, M.; Xiao, N. DP-Pack: Distributed Parallel Packing for FPGAs. In Proceedings of the 2018 International Conference on Field-Programmable Technology (FPT), Naha, Japan, 10–14 December 2018; pp. 282–285. [CrossRef]
13. Maidee, P.; Ababei, C.; Bazargan, K. Timing-driven partitioning-based placement for island style FPGAs. *IEEE Trans. Comput.-Aided Des. Integr. Circuits Syst.* **2005**, *24*, 395–406. [CrossRef]
14. Feng, W. K-way partitioning based packing for FPGA logic blocks without input bandwidth constraint. In Proceedings of the 2012 International Conference on Field-Programmable Technology, Seoul, Republic of Korea, 10–12 December 2012; pp. 8–15. [CrossRef]
15. Feng, W.; Greene, J.; Vorwerk, K.; Pevzner, V.; Kundu, A. Rent's Rule Based FPGA Packing for Routability Optimization. In Proceedings of the 2014 ACM/SIGDA International Symposium on Field-Programmable Gate Arrays, Monterey, CA, USA, 26–28 February 2014; pp. 31–34. [CrossRef]
16. Vercruyce, D.; Vansteenkiste, E.; Stroobandt, D. Runtime-quality tradeoff in partitioning based multithreaded packing. In Proceedings of the 2016 26th International Conference on Field Programmable Logic and Applications (FPL), Lausanne, Switzerland, 29 August–2 September 2016; pp. 1–9. [CrossRef]
17. Arora, A.; Wei, Z.; John, L.K. Hamamu: Specializing FPGAs for ML Applications by Adding Hard Matrix Multiplier Blocks. In Proceedings of the 2020 IEEE 31st International Conference on Application-Specific Systems, Architectures and Processors (ASAP), Manchester, UK, 6–8 July 2020; pp. 53–60. [CrossRef]
18. AN 519: Stratix IV Design Guidelines. Available online: https://www.intel.com/programmable/technical-pdfs/654680.pdf (accessed on 9 June 2023).
19. Arora, A.; Boutros, A.; Rauch, D.; Rajen, A.; Borda, A.; Damghani, S.A.; Mehta, S.; Kate, S.; Patel, P.; Kent, K.B.; et al. Koios: A Deep Learning Benchmark Suite for FPGA Architecture and CAD Research. In Proceedings of the 2021 31st International Conference on Field-Programmable Logic and Applications (FPL), Dresden, Germany, 30 August–3 September 2021; pp. 355–362. [CrossRef]

Disclaimer/Publisher's Note: The statements, opinions and data contained in all publications are solely those of the individual author(s) and contributor(s) and not of MDPI and/or the editor(s). MDPI and/or the editor(s) disclaim responsibility for any injury to people or property resulting from any ideas, methods, instructions or products referred to in the content.

electronics

Editorial

Applications Enabled by FPGA-Based Technology

Andrea Guerrieri *, Andres Upegui and Laurent Gantel

inIT, Hepia, University of Applied Sciences Western Switzerland, 2800 Delémont, Switzerland
* Correspondence: andrea.guerrier@hes-so.ch

Field-programmable gate array (FPGA) technology represents a potential alternative to classical CPUs and GPUs in the post-Moore era from edge computing to data centers. FPGAs offer performance improvements when compared with traditional processing architectures due to their spatial computation capability and energy efficiency. In recent years, FPGA technologies have evolved in different forms of tools, design methodologies, and architectural features. These technologies have enabled or boosted novel application domains. This Special Issue aims to present how advances in FPGA-based technologies have made it possible for multiple application domains.

The following two papers "FPGA-Based High-Throughput Key-Value Store Using Hashing and B-Tree for Securities Trading System" and "Acceleration of Trading System Back End with FPGAs Using High-Level Synthesis Flow" show how FPGA-based technology helps accelerate high-performance trading systems. In first paper, the authors propose a high-throughput key-value store (KVS) for securities trading system applications using an FPGA. The design uses a combination of hashing and B-Tree techniques and supports a large number of keys (40 million), as required by the trading system. The design uses high bandwidth memory (HBM), on-chip memory available in Virtex Ultrascale+ FPGAs, to support a large number of keys. The second paper presents the design of a financial trading system order-processing component using FPGAs, and it was implemented with high-level synthesis (HLS) flow. The order processing component is the major contributor to increased delays and low throughput in the current software implementation of trading systems. The objective of FPGA implementation is to reduce the latency of order processing and increase the throughput of trading systems as compared to software implementation. This paper shows more than double the advantage in order-processing speed and a reduction in latency when using FPGA technology.

FPGA-based applications for machine learning are of a different variety, as shown by "Electromyogram (EMG) Signal Classification Based on Light-Weight Neural Network with FPGAs for Wearable Application", "An Instruction-Driven Batch-Based High-Performance Resource-Efficient LSTM Accelerator on FPGA, and "Improving Seed-Based FPGA Packing with Indirect Connection for Realization of Neural Networks" . In the first paper, to accomplish a lightweight neural network, a maximal overlap discrete wavelet transform (MODWT) and a smoothing technique were used for better feature extractions. Moreover, learning efficiency increased when using an augmentation technique. In designing the neural network, a one-dimensional convolution layer is used to ensure that the neural network is simple and lightweight. Consequently, the lightweight attribute can be achieved, and neural networks can be implemented in edge devices such as the FPGA, yielding low power consumption, high security, fast response times, and high user convenience for wearable applications. The second paper proposes an LSTM accelerator that is driven by a specific instruction set. The accelerator consists of a matrix multiplication unit and a post-processing unit. The matrix multiplication unit uses the staggered timing of read data to reduce register usage. The post-processing unit can complete various calculations with only a small amount of digital signal processing (DSP) slices using resource sharing, and at the same time, the memory footprint is reduced via well-designed data flow

Citation: Guerrieri, A.; Upegui, A.; Gantel, L. Applications Enabled by FPGA-Based Technology. *Electronics* **2023**, *12*, 3302. https://doi.org/10.3390/electronics12153302

Received: 25 July 2023
Accepted: 27 July 2023
Published: 1 August 2023

Copyright: © 2023 by the authors. Licensee MDPI, Basel, Switzerland. This article is an open access article distributed under the terms and conditions of the Creative Commons Attribution (CC BY) license (https://creativecommons.org/licenses/by/4.0/).

designs. The accelerator is batch-based and capable of computing data from multiple users simultaneously. Since the calculation process of LSTM is divided into a sequence of instructions, it is feasible to execute multi-layer LSTM networks and large-scale LSTM networks. Experimental results show that the accelerator can achieve a performance of 2036 GOPS at 16-bit data precision while having higher hardware utilization compared to previous work. The third paper proposed a quantitative rule for packing the priority of neural network circuits and optimized traditional seed-based packing methods using special primitives. The experimental result indicates that the proposed packing method achieves an average decrease of 8.45% in critical path delay compared to the VTR8.0 on Koios deep learning benchmarks.

Furthermore, FPGA-based technology for accelerating algorithms is shown in the following papers. "A Model of Thermally Activated Molecular Transport: Implementation in a Massive FPGA Cluster" reports a massively parallel implementation of Boltzmann's thermally activated molecular transport model. This model allows considering potential energy barriers in molecular simulations and thus modeling thermally activated diffusion processes in liquids. The model is implemented as an extension to the basic dynamic lattice liquid (DLL) algorithm on ARUZ, a massively parallel FPGA-based simulator located at BioNanoPark Lodz. The advantage of this approach is that it does not use any exponentiation operations, minimizing resource usage and allowing one to perform simulations containing up to 4,608,000 nodes. "FPGA Implementation of Shack–Hartmann Wavefront Sensing Using Stream-Based Center of Gravity Method for Centroid Estimation" presents a quick and reconfigurable architecture for Shack–Hartmann wavefront sensing implemented on FPGA devices using a stream-based center of gravity to measure spot displacements. By calculating the center of gravity around each incoming pixel using optimal window matching with respect to the spot size, the common trade-off between noise and bias errors and dynamic range due to window size existing in conventional center of gravity methods is avoided. In addition, the accuracy of centroid estimation is not compromised when the spot moves to or even crosses the sub-aperture boundary, leading to an increased dynamic range. The calculation of the centroid begins when the pixel values are read from an image sensor, and further computations, such as slope and partial wavefront reconstruction, follow immediately as the sub-aperture centroids are ready. The result is a real-time wavefront sensing system with very low latency and high measurement accuracy that is feasible for targeting low-cost FPGA devices. This architecture provides a promising solution that can cope with multiple target objects and work in moderate scintillation.

"FPGA-Flux Proprietary System for Online Detection of Outer Race Faults in Bearings" introduces an online fault detection mechanism for industrial machinery, such as induction motors or their components (e.g., bearings). Most commercial equipment provides general measurements and not a diagnosis. On the other hand, commonly, research studies that focus on fault detection are tested offline or over processors that do not comply with an online diagnosis. In this sense, the present work proposes a system based on a proprietary FPGA platform with several developed IP cores and tools. The FPGA platform and a stray magnetic flux sensor are used for the online detection of faults in the outer race of bearings in induction motors. The integrated parts comprising the monitoring system are the stray magnetic flux triaxial sensor, several developed IP cores, an embedded processor for data processing, and a user interface where the diagnosis is visualized. The system performs the fault diagnosis via a statistical analysis as follows: First, a triaxial sensor measures the stray magnetic flux in the motor's surroundings (this flux will vary as symptoms of the fault). Second, an embedded processor in an FPGA-based proprietary board drives the developed IP cores in calculating statistical features. Third, a set of ranges is defined for the statistical features' values, and it is used to indicate the condition of the bearing in the motor. The results demonstrate that the values of the root mean square (RMS) and kurtosis, extracted from the stray magnetic field from the motor, provide a reliable diagnostic of the analyzed bearing. The platform is based on FPGA XC6SLX45 Spartan 6 of Xilinx, and the architecture of the modules used is described in HDL.

"Finding the Top-K Heavy Hitters in Data Streams: A Reconfigurable Accelerator Based on an FPGA-Optimized Algorithm": This paper presents a novel approach for accelerating the top-k heavy hitter query in data streams using field programmable gate arrays (FPGAs). Current hardware acceleration approaches rely on the direct and strict mapping of software algorithms into hardware, limiting their performance and practicality due to the lack of hardware optimizations at an algorithmic level. The presented approach optimizes a well-known software algorithm by carefully relaxing some of its requirements to allow for the design of a practical and scalable hardware accelerator that outperforms current state-of-the-art accelerators while maintaining near-perfect accuracy. This paper details the design and implementation of an optimized FPGA accelerator that is specifically tailored for computing the top-k heavy hitter query in data streams. The presented accelerator is entirely specified at the C language level and is easily reproducible with high-level synthesis (HLS) tools. Implementation on Intel Arria 10 and Stratix 10 FPGAs using the Intel HLS compiler showed promising results—outperforming prior state-of-the-art accelerators in terms of throughput and features.

Finally, the last two papers presented FPGA-based technology applied to real-time and embedded systems. The first paper, "A New FPGA-Based Task Scheduler for Real-Time Systems" demonstrates a novel design of an FPGA-implemented task scheduler for real-time systems that supports both aperiodic and periodic tasks. The periodic tasks are automatically restarted once their period has expired without any need for software intervention. The proposed scheduler utilizes the earliest deadline first (EDF) algorithm and is optimized for multi-core CPUs that are capable of executing up to four threads simultaneously. The scheduler also provides support for task suspension, resumption, and enabling inter-task synchronization. The design is based on priority queues, which play a crucial role in decision making and time management. Thanks to the hardware acceleration of the scheduler and the hardware implementation of priority queues, it operates in only two clock cycles regardless of the number of tasks in the system. The results of the FPGA synthesis, performed on an Intel FPGA device (Cyclone V family), are presented in the paper. The second paper is "Accurate Multi-Channel QCM Sensor Measurement Enabled by FPGA-Based Embedded System Using GPS" : This paper presents a design and implementation proposal for a real-time frequency measurement system for high-precision, multi-channel quartz crystal microbalance (QCM) sensors using a field programmable gate array (FPGA). The key contribution of this study lies in the integration of a frequency measurement and mass resolution computation based on global positioning system (GPS) signals within a single FPGA chip, utilizing I/O blocks to incorporate logical QCM oscillator circuits. The FPGA design enables parallel processing, ensuring accurate measurements, faster calculations, and reduced hardware complexity by minimizing the need for external components. As a result, a cost-effective and accurate multi-channel sensor system is developed, serving as a reconfigurable standalone measurement platform with communication capabilities. The system is implemented and tested using the FPGA Xilinx Virtex-6, along with multiple QCM sensors. The implementation on a Xilinx XC6VLX240T FPGA achieved a maximum frequency of 324 MHz and consumed a dynamic power of 120 mW. The proposed system meets the precision measurement requirements for QCM sensor applications, exhibiting low measurement errors when monitoring QCM frequencies that range from 1 to 50 MHz with an accuracy of 0.2 ppm and less than 0.1 Hz.

These papers demonstrate the diverse range of applications enabled by FPGA-based technology. As the application requiring high performance and flexibility continues to evolve, we can expect to see even more and more applications in the future.

List of Contributions:

1. Puranik, S.; Barve, M.; Rodi, S.; Patrikar, R. FPGA-Based High-Throughput Key-Value Store Using Hashing and B-Tree for Securities Trading System.
2. Puranik, S.; Barve, M.; Rodi, S.; Patrikar, R. Acceleration of Trading System Back End with FPGAs Using High-Level Synthesis Flow.

3. Jabłoński, G.; Amrozik, P.; Hałagan, K. A Model of Thermally Activated Molecular Transport: Implementation in a Massive FPGA Cluster.
4. Choi, H. Electromyogram (EMG) Signal Classification Based on Light-Weight Neural Network with FPGAs for Wearable Application.
5. Kong, F.; Cegarra Polo, M.; Lambert, A. FPGA Implementation of Shack–Hartmann Wavefront Sensing Using Stream-Based Center of Gravity Method for Centroid Estimation.
6. Mao, N.; Yang, H.; Huang, Z. An Instruction-Driven Batch-Based High-Performance Resource-Efficient LSTM Accelerator on FPGA.
7. Kohútka, L.; Mach, J. A New FPGA-Based Task Scheduler for Real-Time Systems.
8. Cureño-Osornio, J.; Zamudio-Ramirez, I.; Morales-Velazquez, L.; Jaen-Cuellar, A.; Osornio-Rios, R.; Antonino-Daviu, J. FPGA-Flux Proprietary System for Online Detection of Outer Race Faults in Bearings.
9. Ebrahim, A. Finding the Top-K Heavy Hitters in Data Streams: A Reconfigurable Accelerator Based on an FPGA-Optimized Algorithm.
10. Bourennane, A.; Tanougast, C.; Diou, C.; Gorse, J. Accurate Multi-Channel QCM Sensor Measurement Enabled by FPGA-Based Embedded System Using GPS.
11. Yu, L.; Guo, B.; Zhi, T.; Bai, L. Improving Seed-Based FPGA Packing with Indirect Connection for Realization of Neural Networks.

Conflicts of Interest: The authors declare no conflict of interest.

Disclaimer/Publisher's Note: The statements, opinions and data contained in all publications are solely those of the individual author(s) and contributor(s) and not of MDPI and/or the editor(s). MDPI and/or the editor(s) disclaim responsibility for any injury to people or property resulting from any ideas, methods, instructions or products referred to in the content.

MDPI
St. Alban-Anlage 66
4052 Basel
Switzerland
www.mdpi.com

Electronics Editorial Office
E-mail: electronics@mdpi.com
www.mdpi.com/journal/electronics

Disclaimer/Publisher's Note: The statements, opinions and data contained in all publications are solely those of the individual author(s) and contributor(s) and not of MDPI and/or the editor(s). MDPI and/or the editor(s) disclaim responsibility for any injury to people or property resulting from any ideas, methods, instructions or products referred to in the content.

www.ingramcontent.com/pod-product-compliance
Lightning Source LLC
Chambersburg PA
CBHW061520210926
51850CB00008B/1986
9783036587844